According to received historiography, the fundamental issue in eighteenth-century optics was whether light could be understood as the emission of particles or as the motion of waves in a subtle medium. Moreover, the emission theory of light was supposed to have been dominant in the eighteenth century, backed by Newton's *physical* arguments. This picture is enriched and qualified by focussing on the origins, contents, and reception of Leonhard Euler's wave theory of light published in 1746, here studied in depth for the first time. Contrary to what has been assumed, in an important sense, the particle–wave debate only starts with Euler. In addition, Euler's wave theory was the most popular theory in Germany for thirty-five years. Finally, when the emission view of light suddenly became dominant in Germany around 1795, new *chemical* experiments were crucial.

Reflecting on the mathematical, experimental, and metaphysical aspects of physical optics, Casper Hakfoort provides as an epilogue a general picture of early modern science.

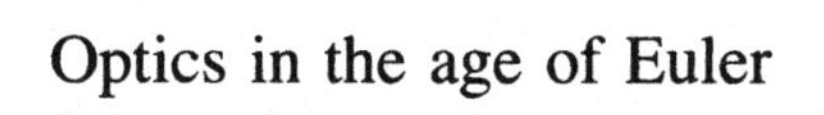

Optics in the age of Euler

Optics in the age of Euler

Conceptions of the nature of light, 1700–1795

CASPER HAKFOORT
University of Twente

CAMBRIDGE UNIVERSITY PRESS
Cambridge, New York, Melbourne, Madrid, Cape Town, Singapore, São Paulo

Cambridge University Press
The Edinburgh Building, Cambridge CB2 2RU, UK

Published in the United States of America by Cambridge University Press, New York

www.cambridge.org
Information on this title: www.cambridge.org/9780521404716

Originally published in Dutch as *Optica in de eeuw van Euler* by Editions Rodopi, Amsterdam, 1986.

First published in English by Cambridge University Press 1995
This digitally printed first paperback version 2006

A catalogue record for this publication is available from the British Library

Library of Congress Cataloguing in Publication data

Hakfoort, Casper.
[Optica in de eeuw van Euler. English]
Optics in the age of Euler : conceptions of the nature of light, 1700–1795 / Casper Hakfoort.
p. cm.
Includes bibliographical references and index.
ISBN 0-521-40471-1
1. Light, Wave theory of – History. 2. Physical optics – History.
3. Euler, Leonhard, 1707–1783 – Knowledge – Optics. I. Title.
QC403.H3513 1995
535′.1′09033–dc20 94-31656
CIP

ISBN-13 978-0-521-40471-6 hardback
ISBN-10 0-521-40471-1 hardback

ISBN-13 978-0-521-03507-1 paperback
ISBN-10 0-521-03507-4 paperback

CONTENTS

ACKNOWLEDGEMENTS

It is a pleasure to thank those who helped to make the present book possible.

This volume is a revised edition of a work first written in Dutch, published in 1986 by Editions Rodopi, Amsterdam. The Netherlands Organization for Scientific Research awarded a translation grant. Enid Perlin-West translated the original Dutch text.

E. A. Fellmann and B. Bosshart provided me with Leonhard Euler's manuscripts. Assistance, intellectual and otherwise, with the Dutch edition was given by H. A. M. Snelders and by C. A. M. de Leeuw, F. H. van Lunteren, L. C. Palm, and C. de Pater. Over the years aspects of the present work were discussed with many people, among them J. Z. Buchwald, H. F. Cohen, M. Feingold, R. W. Home, A. E. Shapiro, and G. N. Cantor. The latter read the English draft and made many valuable comments. None of them will agree with the final text, but I trust that the remaining disagreements will be fruitful.

1

Introduction

The aim of this work is to make a two-fold contribution to the study of eighteenth-century science. The majority of this book is devoted to a description and analysis of the conceptual development of physical optics in the period, focussing on the origins, contents, and reception of Leonhard Euler's wave theory of light. There will always be a second question in the background of the narrative, which will receive full attention in the last chapter: What does a study of eighteenth-century optics have to teach us about the changing nature of natural philosophy and science in that period?

The title of this study – *Optics in the Age of Euler* – constitutes a response to the still generally accepted historical image of optics in which the eighteenth century is portrayed as the century of Newton.[1] According to the standard account, 'Newton's' particle, or emission, theory of light dominated for more than a century, whereas 'Huygens'' wave, or medium, theory supposedly did not develop and found few supporters during the same period. This study provides a corrective to this image, with the Swiss mathematician and natural philosopher Leonhard Euler (1707–83) a leading figure in the new historiographical drama. Euler's importance derives from his "Nova theoria lucis et colorum" (A new theory of light and colours), published in 1746. This article was the foremost eighteenth-century contribution to the development of the medium theories of light. Euler's theory of light, rather than Huygens' theory, was the first serious rival to the emission theories. The present study reveals that Euler's influence was particularly strong in the German lands. Indeed, during the period 1755–90 in Germany the champions of Euler's theory were in the majority.

The title *Optics in the Age of Euler* should not be taken to imply that Euler, rather than Newton, dominated eighteenth-century optics. This is definitely not my intention. However, focussing on Euler and the reception of his theory in Germany allows us to draw a more complete and somewhat different picture.[2] In particular, three themes come to the fore in the present perspective. First, the development of the medium theories of light from Huygens to Euler has been given little treatment in the literature so far. It turns out to be a far more complex and interesting story than has been realized, for example, concerning the different ways in which the concept of periodicity was introduced. Second, studying Euler draws our attention to the slow emergence of the debate between two, and not more than two, optical traditions, the medium tradition and the emission tradition, in the decades before 1746. Contrary to what has been generally assumed, in an important sense the 'wave–particle debate' only starts with Euler. Third, the study of the reception of Euler's theory of light in Germany reveals in a striking way the relevance of chemistry to the assessments of optical theories in the eighteenth century.

In this introductory chapter we shall first examine in greater detail the standard image of eighteenth-century optics and the assumptions underlying it. Next I shall define a number of the main terms used in the text, and finally there is an outline of the way this study is constructed.

1. Condensation and presentism in the historiography of optics

The history of science is sometimes regarded as a triumphal procession of 'great men':[3] The genius, such as Galileo or Newton, produces the insights that determine the course of scientific development. This version of history often leads to 'historical condensation'. Just as the heroic deeds of various people are compressed into the actions of one central figure in the tales of King Arthur, so Newton, the great luminary of the Enlightenment, is superbly suited to the role of legendary hero in the history of science. Consequently, some ideas have been attributed to him, although at best he held them only in an incomplete way or in another form. In many cases natural philosophers coming after him worked out his ideas and suggestions, enriched them with their own discoveries or arranged them in a systematic way. Where their role was forgotten, the image of Newton expanded into an overpowering centre of attraction, rendering the other figures anonymous and encorporating their contributions.

For example, by means of this process the present-day meaning of the second law of mechanics, usually represented by $\mathbf{F} = m\mathbf{a}$, came – in all its glory – to be attributed to Newton. Yet many other scientists, among whom Euler occupies a prominent position, contributed to the history of this law.[4] Newton's role in the history of optics also provides us with a splendid example of historical condensation. We are concerned here with the authorship of 'the' eighteenth-century emission theory of light, which states that light sources emit matter, for example, a stream of immensely tiny particles that produce a visual perception in our eyes. Many writers regarded Newton as the author of this theory, adding in the same breath that the eighteenth century had on the whole accepted the emission theory of light, mainly because of Newton's great authority.[5] This statement not only contains an invalid generalisation embodied in the expression 'on the whole'; it also contains an inadmissable simplification of the actual course of events. Eighteenth-century natural philosophers were in no position to accept Newton's emission *theory*, even had they wanted to do so. Newton's heritage consisted of a number of different, even incompatible, suggestions throughout the *Opticks* and especially in the Queries. The task completed by his contemporaries and successors was not an easy one, and positively demanded a personal contribution from each. Taking Newton's suggestions as their starting point, they constructed a theory demonstrating greater coherence and consistency than the original comments. This 'Newtonian' programme was first completed in the 1720s and 1730s, and it is only from this period that one can talk of the acceptance or rejection of an emission *theory*. Moreover, the 'Newtonian' emission theory *in statu nascendi* was not the only emission hypothesis in the early eighteenth century. There were other embryonic emission theories that bore no relation, or a less exclusive relation, to Newton's suggestions.

According to the received account, Christiaan Huygens was Newton's great rival. He has also suffered historical condensation, for is he not regarded as the foremost advocate of 'the' wave theory of light? However, he cannot be counted as a wave theorist and the main rival of Newton for three significant reasons. First, he explicitly rejected a basic element of a wave theory of light, that is, the periodicity of the pulses.[6] Second, he acknowledged that he had omitted the explanation of colours because the task was too heavy for him. This major omission alone would have been enough to eliminate Huygens' theory as a serious rival to an emission theory. The third reason concerns Huygens' explanation for the rectilinear propagation of light. As will be shown in Chapter 3, this explanation was generally considered inadequate. The alleged failure was cited as the main argument for rejecting Huygens' theory, but Huygens' explanation was also criticized by those who supported a different kind of medium theory.[7] More than half a century after Huygens, Euler produced a medium theory of light that also accounted for a range of colour phenomena. The concept of frequency and the associated periodicity of pulses occupied a central position in Euler's theory. Consequently, only with Euler do we have what we usually call a *wave* theory of light. By contrast, Huygens' conception is best indicated as a *pulse* theory. Although Euler's explanation of rectilinear propagation of light was unable to convince everyone, his theory was considered a serious competitor to the emission theories. (To avoid creating a new legendary figure, it should be mentioned that Euler was indebted to previous optical theorists – among whom Huygens, surprisingly enough, occupies a modest position.)

A peculiarity of many earlier historical studies is that their norms of evaluation were derived from contemporary knowledge. For example, Aristotle's physics or Ptolemy's geocentric world system were not considered from the context of their own times, but were assessed on the basis of a comparison with the current state of physics and astronomy, which often amounted to condemnation or – its complement – praise of insights 'already' possessed in an earlier age. The current day could thus influence the view of the past in this fairly direct manner.

It could also work in a more subtle way: Evaluation was based not on the current state of knowledge, but on the current manner in which science defined its problems. Again, we find an example of this in the historiography of optics. In the 1920s a vital choice to be made in optics lay at the centre of scientific interest: Did light consist of a stream of emitted 'particles' (light quanta), or was it a wave motion? A parallel was drawn with the celebrated debate at the beginning of the nineteenth century, when an emission theory based upon Newton's suggestions and developed further especially in France competed with Fresnel's wave theory for the favours of the scientific community. It was only natural to see the late seventeenth and early eighteenth centuries in the same dualistic terms. Had Newton not defended 'the' emission theory, and Huygens 'the' wave theory? There was a strong temptation to answer in the affirmative here, for what could be more dramatic than a continuous reincarnation of thesis and antithesis, certainly for someone witnessing the synthesis of the contradictions in quantum mechanics?[8]

The dualistic view of the history of optics and the historical condensation where Newton and Huygens are concerned afforded mutual reinforcement. In their close cooperation they created a distorted image of optics at the end of the seventeenth century. According to this picture, the debate between the emission and the wave theories was the central problem at that time, but that view is incorrect, if only because neither theory existed in that period. More important, the optics of the late seventeenth century, as well as that of the early eighteenth century, was dominated by an entirely different issue: the nature of colours. In 1672 Newton had put forward new ideas on this subject, thereby attacking a centuries-old tradition and consequently earning a great deal of criticism from his contemporaries. Admittedly, in the exchanges about colours the opposition between emission and medium conceptions also arose, but the issue had no dominant role and certainly not the elaborately costumed part in which we encounter it seventy-five years later, in the mid-eighteenth century. At that time both emission and wave theories were finally present in a systematically organised form. The ensuing debate did indeed revolve around the question of whether the propagation of light was an emission of matter or a motion in an ethereal medium. However, for the two reasons given here, standard historiography has not recognised the slow emergence of the wave–particle debate in optics in the period between Newton and Euler, and it has missed the significance of the actual wave–particle debate in the second half of the eighteenth century. Ironically, what historians of science on the basis of doubtful premises thought they could find in the seventeenth century, has in fact taken place in the eighteenth century.

2. Terminology and outline of this book

Some of the problems arising from the old historical image of optics are interwoven with the language used. One example of this can be seen in the way some authors talk of 'the' wave theory, making no distinction between Huygens' theory, which was a pulse theory, and later theories of light that could more justifiably be called wave theories. Another example is provided by the identification of Newton's suggestions for an emission theory of light with the systematized emission theories of the mid-eighteenth century. To prevent the elaboration of a new image from being handicapped by these mistaken associations, the key terms used in the following chapters will be discussed.

The first set of terms classifies the various kinds of hypotheses about

Table 1. *Conceptions of the nature of light in Classical Antiquity*

	Effect from the object to the eye	Effect from the eye to the object
Active medium	Propagation of the effect by the medium (Aristotle)	Eye rays with medium (Stoics)
No active medium	Emission of matter (atomists)	Eye rays without medium (Euclid, Ptolemy)

the nature of light. For this I use Table 1, an outline, in a somewhat adapted form, from a recent survey of the history of optics.[9] The table classifies conceptions of light in Classical Antiquity. Viewed schematically, there were four groups of ideas, distinguishable by the answers they provided to two fundamental questions. The first question concerned the way human beings perceive objects: Either the objects have an effect on the eye, or the eye has an effect on the objects. The second basic question concerned the active involvement of a medium between eye and object in the process of seeing. The four 'pure' conceptions consisted of combining pairs of answers. Euclid, Ptolemy, and the Stoics held a theory of 'eye rays', in which the eye scans the objects. For the first two the medium had no part in the process, whereas the Stoics believed that the eye rays were propagated with the aid of the medium, which for them was the air. By contrast, Aristotle assumed that objects acted on the eye via a medium, the 'potentially transparent' that became 'actually transparent' through the action of light-giving objects. Light was a property of the transparent medium, rather than a substance. A different theory was held by the atomists (Leucippus, Epicurus, Lucretius), who posited that an object emitted atoms that constituted a representation of the object. There was no active role for a medium in the propagation of the atoms; on the contrary, the actual medium, the air, was regarded as a disturbing factor in this process.[10]

Theories involving eye rays were largely ignored in the seventeenth century,[11] leaving only two theory groupings: the emission group (with no action from the medium) and the medium group (with objects acting on the eyes via a subtle medium). Pierre Gassendi, who was of great importance for the revival of atomism at the beginning of the seventeenth century, was a representative of the emission group. Harking back to Epicurus, he put forward an emission hypothesis, in contrast to Aristotle's medium hypothesis. In the eighteenth century Gassendi's name was sometimes cited in connection with the emission group, but by then this conception of light had become more closely associated with Newton. René Descartes, who, like Gassendi, held a corpuscular world-view, was the most significant representative of the medium group. Since his name was also cited a great deal in the eighteenth century, the survey in Chapter 3 of the theoretical traditions in physical optics of the period 1700–45 will begin with Descartes' and Newton's contributions.

A great deal of confusion also arises from the use of the term *theory* by historians. The term is used in such different instances as Newton's 'optical theory' in the Queries, Willem Jacob 's Gravesande's theory of light, and 'the emission theory in the eighteenth century'. Taking these in order, we are dealing with a basic hypothesis, with several suggestions for elaborating upon it; a systematically organised theory that is to a certain extent complete; and a theoretical tradition consisting of a succession of various hypotheses and theories with a common basic idea underlying them. In order to clarify the situation, I distinguish among *hypothesis*, *theory*, and *tradition* in the above examples, and I shall do so, too, in the following chapters.

Development within a theoretical tradition is frequently noncumulative. Huygens' pulse theory thus occupies an isolated position within the medium tradition during the eighteenth century, and only at the beginning of the nineteenth century was Fresnel to continue and improve upon Huygens' specific approach. The remainder of the medium tradition in the seventeenth and eighteenth centuries reveals a greater coherence. Descartes' 'pressure hypothesis', Nicolas Malebranche's detailed pressure hypothesis, Johann II Bernoulli's 'fibre theory', and Euler's wave theory have more in common than merely the basic idea of the medium tradition. This is not to say that there were no significant differences of conceptions and approaches. It is precisely in order to emphasize this aspect that these hypotheses and theories are not regarded as examples of one and the same theory, but rather as distinct parts of a theoretical tradition.

A final point concerned with the use of the word *theory*, deserves our consideration: As there is a difference between a description of the phenomena and their causal explanation, so, similarly, there is a difference between a descriptive and an explanatory theory.[12] In the eighteenth century such a distinction was both made and generally accepted. Thus, geometrical optics, the mathematical theory of the passage of rays through mirrors and lenses, was considered to be a theory that provided no role for the physical cause or nature of light: The passage of rays remained the same, whether light was explained by the emission of particles or by motion within a medium. Another theory with descriptive status in the eighteenth century, one that will be discussed several times in this study, is Newton's theory of colours. Newton claimed in 1672 that he had derived the theory solely through experimentation. He therefore believed, somewhat controversially, that his theory of colours was entirely independent of a causal hypothesis on the nature of light. At that time he preferred not to give an explanation of colours based on physical hypotheses. After some initial opposition in the discussion, Newton's epistemological claim was accepted. That meant that his emission hypothesis of light and his theory of colours were regarded as two distinct elements in his optical work: The first – the explanatory hypothesis – could be rejected, for example, whereas the second – the descriptive theory – was accepted.

Following these remarks on terminology, it is easier to give an explanation of the arrangement and the main propositions of this study.

Chapter 2 provides an outline of Newton's new theory of colours and an analysis of the contemporary discussion of this theory. The German reception of his doctrine of colours is then described in greater detail. In the period 1672–c. 1720 the antipathy between the emission and medium traditions was overshadowed by discussion of Newton's theory of colours, which was soon to be separated completely from the debate on the physical explanation of light and colours. The reception accorded to Newton's theory of colours in Germany confirms the latter suggestion.

In Chapter 3 the development of the emission and medium traditions from the end of the seventeenth century to the middle of the eighteenth is outlined. The thought of a number of prominent figures is discussed, and consideration is given to the arguments for each theoretical position and against the opposing positions. This chapter includes the thesis that the opposition between the emission and medium traditions only increased slowly, and that only with Euler, in 1746, did it acquire the form that present historiography considers to have existed some seventy-five years before that date.

Chapter 4 is entirely devoted to Euler's wave theory. First of all, his theory is located within the medium tradition. Next comes a discussion of Euler's general arguments for and against emission and medium conceptions of light. Finally, some of his fundamental theoretical points are analysed in detail. It is suggested in this chapter that Euler was the first optical theorist to put forward a fairly complete and influential *wave* theory, that is, a medium theory of light including periodical waves; that in doing this he was, in a very general sense, building upon foundations that Huygens had laid, while at the same time rejecting some of Huygens' basic ideas; that the cool reception accorded his wave theory by his Berlin colleagues resulted in the influential first chapter of Euler's principal work on physical optics (the 'Nova theoria lucis et colorum'), in which he drove home the antipathy between the emission and medium traditions through a comprehensive analysis of both; that Euler's greatest conceptual contribution lay in the anchoring of the concept of frequency in the medium tradition, and that the related climax of this theory was found in the hypothesis – later disproved – of the existence of 'undercolours'. (Euler held that there were colours with frequencies higher and lower than those in the solar spectrum; I have chosen to name them overcolours and undercolours, by analogy with overtones in music.)

Chapter 5 treats of the German debate on the nature of light that followed the publication of Euler's wave theory. During most of this debate, which lasted for about fifty years, the supporters of a medium theory outnumbered the defenders of an emission theory. By way of introduction there is an outline of the situation circa 1740 and of the support enjoyed by the different light theories in the subsequent half century. Description and analysis of the content of the debate comprises the main part of the chapter. Included are two early contributions to the discussion, in which Euler's general arguments were dealt with one by one. As in the rest of the debate, the details of his wave theory were not taken into account. The first contribution repudiated Euler's line of argument, whereas the second accepted it. Clearly Euler's arguments were unable definitively to conclude the debate on light. A detailed treatment of the discussion of two physical phenomena follows. The first phenomenon is the rectilinear propagation of light, for the major objection to the medium theories was that their explanation of this phenomenon was incorrect. The theoretical ideas and new experiments advanced in connection with this aspect of the discussion failed to close the debate on light. The same applies to the second physical phenomenon considered in the debate: the experiment on 'grazing light rays' (rays that travel close to the surface of glass, for example), which played a modest but intriguing role in the German discussion. The debate changed its character in the mid-1770s, when interest was aroused in a phenomenon with both physical and chemical aspects: phosphorescence. Experiments on the absorption and production of coloured light by phosphorescent substances stimulated a discussion in which Euler participated. However, the experiments on phosphorescence could not end the debate. The debate only concluded after around 1789, when the chemical actions of light (the formation of oxygen by green foliage and the blackening of silver salts, for example), which had been extensively investigated shortly before, were added to the discussion agenda. The conclusion was unexpectedly rapid and complete. The medium tradition lost its dominance within a few years and was replaced by the emission tradition. Since Germany joined in the international pattern of the history of optics around 1795, that year provides a natural end-point for this study.

Chapter 6 constitutes an epilogue, although not in the usual sense, as it contains no outline of developments within the field of optics after 1795. It is presented as an epilogue in the *historiographical* sense. Set against the background of a number of conclusions from the previous chapters, a historiography of science for the seventeenth and eighteenth centuries is outlined. This suggestion draws on the work of Thomas Kuhn on the historical development of the physical sciences, but it also signals a fundamental change to his schema.

2

The debate on colours, 1672–1720

1. Newton's theory of colours

a. The modification theory of colours

Why is one object red and another blue? Aristotle believed colours to be a mixture of light and darkness. In his view an object is white when all the light striking it is reflected, without the addition of any darkness, and an object is black because it reflects none of the light falling upon it. The colours of objects derive from the mingling of light and darkness in varying proportions. Darkness may originate in something opaque or, as in the case of the rainbow, in an opaque medium, such as the clouds. Red, the purest colour, is a mixture of light and a small amount of darkness. As the amount of darkness increases, green is observed and eventually violet, the 'darkest' colour. The other colours consist of a combination of red, green, and violet, the three primary hues. It is fundamental to this interpretation that colours are a *modification* of pure and homogeneous white light, resulting from the addition of darkness.[1]

The modification theory of colours, which – like so many of Aristotle's ideas – seemed to fit so well with direct observation, was generally accepted until the second half of the seventeenth century, although with variations. Some writers assumed the existence of two or of four primary colours; others opted for three, but chose different hues than Aristotle's. For example, Athanasius Kircher selected yellow, red, and blue.[2] Descartes adopted the basic principles of Aristotle's theory, a fact that will perhaps surprise those who regard Descartes as having made a fundamental break with the theories of Aristotle and scholasticism. There is continuity with tradition even in the most innovative aspects of his philosophy.[3] However, Descartes did change the modification theory, reinterpreting it in mechanistic terms. Mechanicism constituted an important, if not the most important, guiding concept for Descartes and many other seventeenth-century natural philosophers. Descartes' mechanicism demanded that the explanation of a natural phenomenon be restricted to the use of particles of matter varying solely in size, form, or movement. In short, the world was seen as consisting of particles differing in a quantitative way, such concepts as quality, sympathy, and force being taboo.

In line with his mechanicism, Descartes conceived light as the action of a subtle matter or ether filling the whole of space. The sun or another light source within this matter produced a 'tendency to motion' among the extremely small particles that we could – somewhat anachronistically – call a pressure. The inclination towards movement is transmitted along straight lines in the ether, the rays of light.[4] Colours result from a special motion of the particles: Descartes explains the coloured edges of a beam of white sunlight passing through a glass prism by positing that the particles show both a forwards and a rotatory motion. (He seems to be inconsistent here in his use of actual movement versus a tendency to move. This will be discussed in the next chapter, as it has no great significance at this point.) The ether particles in the beam of light rotate at a velocity identical to the speed of translation, whereas particles in the shadow do not rotate at all. At the borders of the beam the stationary particles retard or accelerate the moving ether particles. The speed at which these particles rotate determines the colour, the particles that turn most rapidly producing red, whereas blue is produced by the particles with the least rotatory motion.[5]

There is a striking resemblance to Aristotle, since Descartes likewise considered colour to be a modification of light influenced by shadow or darkness. Descartes translated the variation in the 'strength' of colours by means of a mechanistic model in which the decisive factor was the rotatory movement. It is ironic to see Descartes in fact lending respectability to Aristotle's views by interpreting them mechanistically, though it may be doubted whether he or his contemporaries were aware of this irony. Many seventeenth-century natural philosophers endorsed the mechanistic modification theory, although not all of them adopted Descartes' model.[6] The strength of the powerful alliance between Aristotle's modification theory and seventeenth-century mechanicism can be gauged from reactions to an article published in 1672 by a young and as yet unknown Englishman.

b. Newton, a dissident

On 6 February 1672 the twenty-nine-year-old Isaac Newton wrote to Henry Oldenburg, secretary of the Royal Society. Two days later his letter was read at the Society's meeting, and it was published shortly after this in the *Philosophical Transactions*. It became known under the title 'New theory about light and colors'.[7] In this letter, Newton's first article, he argued for a new theory of colours, reporting on the optical experiments he had done several years earlier.

Newton's theory is in complete contrast to the modification theory of colours. Newton maintained that white light is a heterogeneous mixture of the colours of the rainbow, rather than the pure and unmixed light it was generally believed to be. These colours are not formed in the process of refraction, since they are already present in white light. No modification of light occurs; on the contrary, impure white light is separated into its constituent parts, the pure and unmixed colours of the spectrum.

The article evoked a great many responses, and in order to understand them, it is necessary to examine the way that Newton presented his most significant conclusion. The first part of the article is introduced as a chronological report on his research. We should not take this literally; it is in fact a logically constructed argument intended to convince the reader of the validity of his theory.[8] The beginning of the investigation, in the logical sense, is a remarkable observation made with the aid of a glass prism. Newton darkened his room and caused light to enter through a circular hole in the window shutter. Through a prism the light formed a colour spectrum on the wall. Earlier researchers, for instance, Descartes and Robert Hooke, had been primarily interested in qualitative problems such as the origins of colours and the physical explanation for them.[9] In contrast, Newton's attention was directed towards a quantitative factor, the shape of the spectrum. To his amazement he found that the colours are not contained within a circular shape. He observed that the spectrum is elongated in form, straight at the edges and rounded at the ends. The length of the colour band is approximately five times larger than the width.[10]

An isolated observation cannot produce anything surprising. What is observed is only extraordinary or deviant in relation to a specific expectation. In the case in hand the law of refraction provided the necessary background. According to this law, first published in 1637, the sines of the angle of incidence and of the angle of refraction are in the same proportion as a number specific for the refracting medium: $\sin i/\sin r = n$.[11] The implicit assumption was that this law applied to 'light', that is, to white light. Colours had no special role to play, being regarded a secondary effect on the borders between light and shadow. In modern terminology the index of refraction n was thought to be independent of the colour of light.

Consequently, according to the law of refraction as it was understood in Newton's time, the spectrum ought to be circular rather than elongated, that is, if the condition is fulfilled that the prism is placed in a special position, that of the 'minimum deviation'. If the prism is not in this position, the law of refraction predicts an elongated spectrum. Newton does not state whether his prism is in the minimum-deviation position, an omission that created understandable confusion among his critics.

Seeking the cause for the distinctive shape of the spectrum, Newton adopted – provisionally – the law of refraction for white light. He discusses several hypotheses to explain the elongated form, given this law. The notable difference between length and width cannot be rendered comprehensible by any of these hypotheses.[12] The repudiation of his 'suspicions' ultimately led him to his famous *experimentum crucis*. The experiment is described verbally in the article. Only later, when asked for elucidation, did he produce a somewhat imprecise drawing of the way in which he had set up his experiment (see Figure 1).[13] The arrangement comprised two prisms and two pieces of card. Sunlight, S, entered through a small hole in a window shutter. A beam of light was refracted in the first prism, subsequently passing through a small opening in a screen placed directly behind the prism. The second screen lay approximately three metres away from the first. Part of the light went through a hole in this, to be refracted for the second time in a prism, before reaching the wall. During the experiment the prism close to the window was slowly turned round on a horizontal axis, enabling the different parts of the spectrum to pass consecutively through the opening in the first screen. Newton observed the places at which the rays thus selected reached the wall. He ascertained that light on the violet side of the spectrum was considerably more refracted by the second prism than was light on the other, red side. In this way, the real cause of the form of the spectrum was revealed to Newton: '*Light* consists of *Rays differently refrangible*, which . . . were, according to their degrees of refrangibility, transmitted towards divers parts of the wall'.[14]

The refraction index consequently depends not only on the kind of

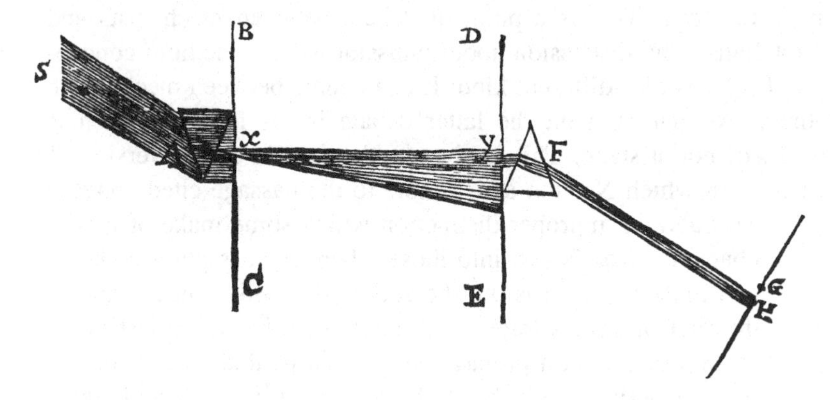

Figure 1. Newton's *experimentum crucis* (Newton, ed. Cohen, *Papers*, 101).

medium but also on the colour of the light. Moreover, colours are not modifications of white light ('*Qualifications of Light*') arising from refraction. Rather, they are '*Original and connate properties*' and, Newton adds, unchangeable attributes of rays of light. However, the greatest surprise of all is the 'wonderful composition' of white light, which is composed of the prismatic colours mixed in the right proportion.[15]

Only at the end of his article does Newton allow the reader to see something of his concept of the nature of light, kept in the background up to that point. This concept emerges when he writes about the colours of opaque bodies, explaining that colours originate in the selective reflection of light rays. A red body is red because it reflects mainly red rays. In light of a different colour the red object shows only a little of that different colour; it thus reflects only a small quantity of those rays. Newton ends the discussion of the colours of opaque bodies with the following words:

> . . . it can be no longer disputed, whether there be colours in the dark, nor whether they be the qualities of the objects we see, no nor perhaps, whether Light be a Body. For, since Colours are the *qualities* of Light, having its Rays for their intire and immediate subject, how can we think those Rays *qualities* also, unless one quality may be the subject of and sustain another; which in effect is to call it *substance*.[16]

It is tempting to read in this passage that Newton assumes that light consists in the emission of matter and that he rejects the conception of light as a motion in a medium. However, his real problem is whether light is material. This is a point of debate between mechanists and Aristotelians. The discussion about emission versus medium conceptions of light is of a different kind: It is a debate between mechanistic brothers. At that moment the latter debate is, as far as Newton is concerned, not at stake. This is abundantly clear in a later version of the paper, in which Newton adds a note to the passage cited above:

> Through an improper distinction which some make of mechanical Hypotheses, into those where light is put a body and those where it is put the action of a body, understanding the first of bodies trajected through a medium, the last of motion or pression propagated through it, this place [i.e., 'no nor perhaps, whether light be a Body': CH] may be by some unwarily understood of the former: Whereas light is equally a body or the action of a body in both cases. . . . Now in this place my design being to touch upon the notion of the Peripateticks, I took not body in opposition to motion as in the said distinction, but in opposition to a Peripatetick quality, stating the question between the Peripateticks and Mechanick Philosophy by inquiring whether light be a quality or body.[17]

2. Reception of Newton's theory of colours

a. Initial reactions

The first reactions to Newton's 'New theory about light and colors' were many and critical. Over a six-year period more than twenty letters were exchanged, the majority of them printed in the *Philosophical Transactions*. The most notable participants in this discussion were Ignace Gaston Pardies, Robert Hooke, and Huygens. Several issues can be distinguished in their criticisms.

To begin with, there was the vagueness of Newton's experiments, the description of the *experimentum crucis* in particular being obscure in the extreme. Pardies was able to understand how the land lay only after receiving two clarifications from Newton, including the sketch of the experimental set-up (Figure 1). Huygens complained about Newton's unclear presentation of his experiment.[18] This lack of clarity made it difficult at first for Newton's opponents to accept his radical conclusions, which were largely based upon the *experimentum crucis*.

Newton and his critics also interpreted Newton's words differently.

According to the hypothesis ascribed to Newton by Hooke and Pardies, white light consists of an infinite variety of light particles, each of which has an immutable colour, comparable to a mixture of extremely fine-grained paint powder flung into space from light-producing objects.[19] In contrast to this supposed emission of matter, which was held to be fundamental to Newton's theory of colours, they posited their own hypotheses of a motion in the ether, colours being caused by a 'disturbance' or 'diffusion' of the ether's motions.[20]

Finally, although no one expressed the perception in so many words, one has the impression that Newton's critics, especially Hooke, perceived in his alleged emission hypothesis the assumption that immutable particles are distinguished by *qualitative* properties, of which colour is one. Such a conception of matter runs counter to the dogma of mechanicism, which stipulates that secondary qualities are not fundamental, being only sensory effects of quantitative properties of matter, such as size, form, and motion, which are considered the primary qualities.[21] Even when this heretical concept of colours was not attributed to Newton, there still remained much to be desired for an orthodox mechanist. Newton's theory of colours lacked the most important feature, a physical model in which colours were reduced to the basic mechanistic categories; as Huygens put it, Newton had not shown how the diversity of colours could be explained by 'la physique mechanique'.[22]

Newton answered that he had not intended to produce an uncertain hypothesis on the nature of light and colours. The 'perhaps' in the penultimate quotation of Section 1 has to be viewed as a sign that Newton himself did not regard the material, mechanistic interpretation as an indispensible basis for his theory of colours. At best this was a plausible conclusion consistent with his theory. Moreover, he claimed that the theory of colours he presented in his article was based exclusively on experiments and was not linked to any explanatory hypothesis. When he talked of colours as qualities of light, it was only in a manner of speaking: Colours still had to be mechanically explained.[23] He even went so far as to offer Hooke suggestions for the completion of the latter's medium theory of light. One of his interesting hints was the linking of colours with the 'various depths or bignesses' of the vibrations in the ether, analogous to the variety of vibrations in air, producing different tones. Newton was probably thinking of wavelengths rather than of amplitudes.[24] Hooke did not follow up on this suggestion, but the analogy between sound and light continued to be frequently made, ensuring that the concept of periodicity would resurface.

The long-drawn-out discussion in the *Philosophical Transactions* had not been in vain. It had cleared up misunderstandings about Newton's experimental arrangement and his conclusions. According to the critics, he had reduced his claims by acknowledging that he had not produced an explanation for colours, but only a descriptive theory. Where this descriptive theory was concerned, Newton emerged from this debate as the winner: In the end his experimental conclusions seemed to be accepted as valid.

Three years after the end of the first debate a critique of Newton's theory of colours appeared, in which Newton was attacked with his own weapons. In 1681 the French experimentalist Edmé Mariotte contested one of Newton's main conclusions, the immutability of the spectral colours, on the basis of a repetition of the *experimentum crucis*.[25] Mariotte's critique is embodied in a broad-based treatise on colours. While discussing the colours of the prism, he addresses several explanatory hypotheses. He devotes considerable attention to Descartes' rotation hypothesis, rejecting it partly because, if the hypothesis were correct, colours would also be produced in the process of reflection, which is not the case. He also dismisses other people's ideas that colour originates in the 'dilution' and 'thickening' of light, since red rays remain red regardless of the distance at which one receives them on a screen, despite the dilution that occurs in a diverging beam. Mariotte called Newton's theory of colours 'une hypothèse nouvelle & fort surprenante'. It is remarkable that the French experimentalist, who later on confined his attention to providing some 'principes d'expérience', classed Newton's theory of colours – which according to Newton himself, was deduced from his experiments and contained no explanatory hypothesis – with the mechanistic hypotheses of Descartes and others. Here, Newton's methodological distinction between descriptive theories and explanatory hypotheses was either not understood or not accepted.[26] Whatever may have been the case, Mariotte was only disputing Newton's conclusions by repeating the latter's experiments, and he therefore in fact treated the new theory of colours as a descriptive theory.

At first Mariotte praised Newton's theory of colours, thinking that the majority of observations and experiments could easily be explained with its help. However, he believed that there were also several experiments that did not accord with Newton's theory. He described one of these experiments at length; this was a repetition of the *experimentum crucis*. He made white light fall on a prism, then allowed the extreme portion of the violet light to be refracted in a second prism. According to Newton there ought to be no change in the colour of the light, but Mariotte found red and yellow – colours from the opposite side of the spectrum – in the newly refracted violet light.[27] Mariotte claimed that violet light clearly suffers a modification in the second prism, being partly transformed into red and yellow. Consequently 'l'ingénieuse hypothèse de Monsieur *Newton*' could not be accepted.[28]

Mariotte enjoyed great authority in matters concerning physical experiments, and it was partly for this reason that his criticism of the *experimentum crucis* prevented Newton's theory of colours from being accepted in France for a long time.[29] Only with the publication of the *Opticks* in 1704 did this situation change. The Académie des Sciences devoted much attention to Newton's work, copious extracts from the *Opticks* being read at nine sessions between August 1706 and June 1707.[30] As we shall see in the next chapter, the influential academician Nicolas Malebranche became convinced that Newton's theory of colours was correct. This conviction came from these sessions and from his own reading of the Latin translation of the *Opticks* published in 1706. According to Guerlac's study of the subject, this did not mean that France in general was won over to the new theory of colours. First, John Theophilus Desaguliers had to repeat some of Newton's colour experiments successfully in front of Royal Society members in 1714 and 1715, a handful of Frenchmen being in the audience in 1715. Several repetitions of the experiments performed on French soil also played a part in encouraging the acceptance of Newton's ideas. In Guerlac's view the two French editions of the *Opticks* appearing in 1720 and 1722 signalled the general acceptance in France of Newton's theory of colours.[31]

b. Reception in Germany

Until recently the secondary literature conveyed the impression that Newton's theory of colours encountered a great deal of resistance in Continental Europe, not only immediately after it was first published in 1672 but also well into the eighteenth century. Guerlac has corrected this account in relation to France, and I shall now provide a similar modification for the German-speaking areas, where the new theory of colours was accepted even more rapidly than it was in France.

Newton's theory of colours met with little response in Germany before the *Opticks* was published in 1704. Although Gottfried Wilhelm Leibniz knew the theory and was interested in it, as one can gather from his correspondence,[32] his familiarity did not result in a wider acquaintance with Newton's ideas, let alone their acceptance. As far as I know, the only German publication of this period in which the topic was discussed is a treatise dating from 1694 attributed to Martin Knorr, a professor of mathematics at Wittenberg.[33] At the end of his discourse on the refraction of light Knorr provides a short summary of Newton's theory of colours, and of Mariotte's objections to it. He pronounces no judgement on the new theory.[34] One can deduce from his account that he entertained reservations about Newton's theory, particularly in the light of Mariotte's criticism, but it would be going too far to interpret Knorr's viewpoint as an outright rejection of the theory. One can say the same for Leibniz, who regarded the new theory of colours as very important but did not want to ignore Mariotte's objections.[35] Leibniz adopted a more positive stance with the publication of the *Opticks* in 1704. Clearly impressed by the numerous meticulous experiments, he wrote that he was prepared to accept the new theory on a provisional basis, despite Mariotte's objections, in the expectation of new trials; he wrote this in an unpublished manuscript that can be dated to shortly after 1704.[36] However, Leibniz was never an uncritical champion of Newton's theory of colours; in a letter written in 1708 he urged a careful and unbiased repetition of Newton's experiments.[37]

Publication of the *Opticks* did create a clear preference among others for the new theory of colours. The most important of these was Christian Wolff, who exercised a considerable influence over the world of German philosophy and science through his textbooks. It is impossible to deduce Wolff's attitude from his summary of the *Opticks* (1704) that appeared in the *Acta eruditorum* of 1706, since his discussion of Newton's theory is couched in the neutral tones characteristic of the *Acta*.[38] In contrast, he champions Newton's theory of colours in the first German edition of his textbook on mathematics, published in 1710. Relating that the Englishman performed a great number of experiments, Wolff provides brief details of three, including the *experimentum crucis*, after a summary of the principal features of Newton's theory.[39] As far as I know, Wolff was thus the first on the Continent to accept Newton's theory of colours in a printed work, predating by two years Malebranche's published support for the theory, which up to the present has been regarded as being the first (see next chapter).

It is unclear whether Wolff was aware in 1710 of Mariotte's critique of Newton. Although he mentions Mariotte's tract on colours in the annotated reading list in his textbook, he makes no mention of Mariotte's objections to Newton's *experimentum crucis*.[40] Wolff had obviously been convinced of the accuracy of Newton's theory by its presentation in the *Opticks*. Nevertheless, Wolff was familiar with Mariotte's disagreement by 1713 at the latest, as one can see from an anonymous contribution to the *Acta eruditorum* of that year, Wolff being the most likely author. A text of some significance, it has been incorrectly interpreted in the historical literature. It contains a discussion of Jacques Rohault's textbook on physics, in Samuel Clarke's Latin edition. Rohault provides a 'Cartesian' treatment of light and colour, but Clarke contrasts this with Newton's ideas in his extensive annotations. The reviewer in the *Acta* provides the following comment on Newton's theory of colours:

> The extremely sagacious Mr. Newton has very successfully refuted the objections which learned men in both France and England voiced against this theory, as appears clearly enough in the Philosophical Transactions . . . ; many therefore hope that he will condescend to devote attention to the problem which has been raised about this theory in a not unhappy manner by the very talented Mariotte, a naturalist indefatigable and not unfruitful in his time, in his treatise on colours. . . .

Then follows a short description of Mariotte's experiments in which violet light appeared to change partly into red and yellow, and red light partly into blue and violet (see above, Section 1b):

> Now, if this change [i.e., of the colours] is acknowledged it is clear that the Newtonian theory . . . collapses. Mariotte also took a distance of 30 feet [between the first and second prism] so that no one could raise the objection that the complete separation of the heterogeneous rays was not achieved yet at a smaller distance. In our view Mariotte's experiment is decisive only if all blue [i.e., violet] light had been changed into something else.[41]

In two respects this text has been viewed in the wrong light. First, many commentators, perhaps following Joseph Priestley's lead, attribute the discussion of Rohault's book to Leibniz.[42] This is incorrect for the review as a whole. The review was Wolff's, although it is certainly not impossible that Leibniz made changes in or additions to the text, since he was still Wolff's intellectual mentor at that time.[43] In view of Leibniz' interest in the experimental testing of Newton's theory of colours, it is conceivable that the passage cited, not the review, ought to be partly or wholly attributed to Leibniz. Since Wolff included a passage strongly resembling the fragment of 1713 in the Latin edition of his textbook on mathematics, published in 1715, he obviously adopted the text quoted above as his own viewpoint, whether or not he conceived it himself in the first place.[44]

Second, the purport of the passage from 1713 is controversial. Newton definitely regarded the observation in the *Acta* as a challenge from what he saw as 'the Continent', and he asked Desaguliers to repeat his experiments with colours. This was done in 1714 and 1715 (see Section 2a), a report on them appearing in the *Philosophical Transactions* of 1716.[45] In line with Newton's view, some historians have created the impression that the remark in the *Acta* expressed a general rejection of Newton's theory in the German-speaking world.[46] This does not accord with the facts. Wolff had already approved the theory of colours in 1710, and a second German-speaking author was to follow suit a year later, as we shall see. Moreover, assuming that the 1713 passage was written by Wolff, the appeal to answer Mariotte's criticism can be interpreted as an attempt by Wolff to dispose of doubts about the theory of colours entertained by others. If Leibniz wrote the passage, the matter is somewhat less clear-cut. Nevertheless, given the manuscript mentioned above, dating from the period soon after 1704, it is clear that Leibniz did not oppose the new theory. After all, he preferred Newton's views to those of Mariotte, in the absence of any further experimental data. If Leibniz was the author of the passage in the *Acta*, the article can be viewed as an attempt to dispel his own reservations, which, after all, were justified in view of the scientific norms of his time.

Thus, although there is still some room for differences in interpretation of the 1713 passage, one thing is beyond dispute: Newton's theory of colours began to be accepted relatively early in Germany, influenced by the *Opticks* (1704) and the *Optice* (1706), even before Desaguliers' repetitions of Newton's experiments in 1714 and 1715.

Wolff continued his active promotion of the new theory of colours after publication of the Latin edition of his textbook on mathematics in 1715. He wrote an article in the *Acta eruditorum* in 1717, in response to the publication of Desaguliers' repetition of Newton's experiments. The confirmation of Newton's theory is mentioned with approbation. In line with Desaguliers, Mariotte's divergent results are ascribed to an incomplete separation of the rays, and consequently the correct method for achieving this separation is copied fairly extensively from Newton's *Opticks*.[47] The new theory of colours also found its way into Wolff's textbook on physics, published in the early 1720s. The volumes on experimental physics, published in 1721–3, included a substantial section on optical experiments. In the treatment of Newton's *experimentum crucis* Mariotte's objections were cited and forcefully repudiated, the reader being referred to Wolff's own ideas of 1715 on this subject and to the 1716 publication of Desaguliers' repetition of Newton's experiments. In his textbook on theoretical physics, published in 1723, Wolff once again applauded the new theory of colours.[48]

In view of the great influence of Wolff's textbooks,[49] he can be regarded as the chief populariser of Newton's theory of colours in the German lands, but there was a second advocate who, although enjoying less influence than Wolff, should not pass unnoticed. He was the Swiss doctor and naturalist Johann Jakob Scheuchzer. In his case, too, the conviction created by the *Opticks* was responsible for the adoption of the new theory of colours.

Scheuchzer published *Physica, oder Natur-Wissenschaft* in 1701; it was reprinted four times up to 1743.[50] The work is designed in a compendium-like way, providing a great many ideas formulated by other authors, but with little personal commentary. In discussing the nature of light in 1701 Scheuchzer opts unequivocally for Descartes' ideas and thus for a medium hypothesis; on the subject of colours he expresses no preference for any particular author, which amounts to de facto support for a concept of modification. His viewpoint on the nature of light is the same in the third edition of 1711.[51] However, there is a drastic change as regards the theory of colours. He has not only opted for Newton's new theory; he is also so enthusiastic that he devotes a whole chapter to an extensive account of the *Optice* of 1706. He regards this as 'ein Meisterstuck des menschlichen Verstands' that 'in einer so dunklen und schweren Sach, wie die gegenwertige ist, so vil Liecht gibet, das wir erst jetzt anfangen die Augen aufthun'.[52] With this, Scheuchzer becomes the second German-speaking author to accept Newton's theory of colours in a printed work, even before Malebranche's and Desaguliers' publications.

It is difficult to determine, with any precision, the route by which

Newton's theory of colours reached other German writers. Johann Wentzel Kaschube provides an example. In 1718 he cites Wolff as his principal mentor, yet he also refers the reader, in a note, to Scheuchzer's summary of Newton's *Optice*.[53] In this note and elsewhere he refers to passages in the *Optice* itself. Although this does not necessarily mean that he had actually read Newton's book, it does remind us of the possibility that, as in the case of Wolff and Scheuchzer, Newton's work exercised a direct influence. Furthermore, other accounts of Newton's theory of colours, such as the Rohault-Clarke textbook of 1710 or the textbooks by 's Gravesande and van Musschenbroek published in 1721 and 1726, may have provided information on the new theory. This does not detract from the fact that the early acceptance of Newton's theory of colours in German textbooks has to be located in the volumes by Wolff and Scheuchzer, published in 1710 and 1711, respectively, and that in each case Newton's work spoke for itself. Remarkably enough the early adherents to Newton's theory of colours did not accept his emission hypothesis on the nature of light. As we have seen, both Wolff and Scheuchzer supported a medium conception of light.

German opposition to Newton's theory of colours was not lacking, but there was not very much of it, and it had little influence. One example is a treatise published in 1720 by Johann Friedrich Weidler, professor of mathematics at Wittenberg.[54] Weidler attempts to adapt Newton's experiments to the old modification theory, providing a commentary based on the literature, rather than a report on new testing. First, the most important ideas and experiments are mentioned, including the criticisms by Mariotte and the repetition of Newton's experiments by Desaguliers.[55] It is clear that Weidler is aware of recent publications, since he provides long quotations in English from Newton and Desaguliers. Nevertheless, as we shall see shortly, Weidler appears to have read the literature, and especially Newton's work, without any great care. He subsequently expounds his own conceptions. Colours, he claims, are light weakened by darkness. He illustrates this by the experience Mariotte had also mentioned, in which the colours appearing behind our closed eyelids after looking into the sun are first red, then yellow, blue, grey, and finally black.[56]

The remainder of Weidler's treatise is devoted to an extensive discussion of Newton's experiments and conclusions. Weidler's commentary is qualitative, containing no calculations and paying no attention to quantitative data, and his own ideas about colours remain vague. He believes that the 'strength' of light determines its refraction and colour, providing by way of example the fact that the further a light source is from a convergent lens, the greater the refraction of the rays of light will be, with a shorter focal length. He attributes this to the greater 'weakness' or 'thinness' of the rays. He then links 'strength' with colour, and – along with Newton – colour with refraction. This brings us full circle: Because the rays of the 'darker' colours are 'weaker', they have a greater refraction.[57] By creating a link between 'strength' and the colour of light, Weidler is adhering to the old theory in which colours come into existence through a modification of white light, by thinning or weakening them with darkness. At the same time he inserts Newton's concept of refraction into the old theory, in a superficial way.[58]

Weidler's tract had little influence; where it was mentioned it was only to be rejected unambiguously. For example, in the article on colours in Zedler's encyclopaedia (1735), only one of the seventeen columns is devoted to Weidler's treatise, principally to indicate that the admixture of darkness can affect only the intensity of the light, not its colour. In 1736 Weidler himself also adopted Newton's theory of colours, without reservations.[59]

Not only was there little opposition in the German lands, but Newton's theory of colours was even defended there against critical attacks by others. Thus, Georg Friedrich Richter, who was to become extraordinary professor of mathematics at Leipzig in 1726, defended Newton against the Italian count Giovanni Rizzetti. Rizzetti claimed in an article in the *Supplementa* to the *Acta eruditorum* (c. 1722) that he had duplicated all of Newton's experiments and had found some of them to be inaccurate. Moreover, because Newton had failed to mention significant details in performing the tests, the remaining results were ambiguous and thus less suitable for supporting Newton's theory.[60] One of the experiments that Rizzetti claimed were incorrect was a test with a piece of paper, one half coloured red, the other, blue. Black thread was wound round the coloured areas, which were illuminated by a candle. The reflected rays travelled through a lens, forming an image of the object upon a white screen. In Newton's view there was a marked difference in focal distance, and therefore in refraction, between the red and blue rays: the blue surface was clearly visible at a smaller distance from the lens than was the red (the sharpness of the image could be precisely determined by observing the image of the black thread).[61] Rizzetti claimed that he had been unable to obtain this result in spite of all his efforts: In his trials the two colour surfaces were equally sharp or blurred.[62] Partly on these grounds he contested the different refrangibility of the colours (one of the main propositions in Newton's theory). Richter reacted to Rizzetti's claim by expressing his firm conviction in the validity of Newton's experiments, reminding his readers of Desaguliers' repetitions of these before an Anglo/French audience. In Richter's view nothing was more widely recognised and more convincingly demonstrated than Newton's experiments.[63] In his response he provided more detailed criticism without actually repeating any of the experiments himself. These were performed in England, where in the meantime Rizzetti's critique had attracted attention. In December 1722 Desaguliers successfully repeated the two-colour experiment in front of the Royal Society[64] at the request of Newton, who himself wrote three versions of a reply to Rizzetti, none of which was published.[65]

The dispute between Rizzetti and Richter dragged on until 1724, in which year Richter declared that as far as he was concerned the matter was closed.[66] Rizzetti for his part continued the dispute, publishing a book in 1727 containing his collected objections to Newton's theory of colours.[67] Desaguliers seized upon this work with a fierce criticism of Rizzetti's methods and of his scientific pretensions, charging him with adopting an arrogant attitude towards Newton and Richter. He illustrated this allegation with pages of citations in which Rizzetti had, he felt, flouted the norms of scientific propriety.[68] Finally, after providing a short account of Rizzetti's book Desaguliers reported on a series of experiments confirming Newton's theory of colours, which had been performed in the presence of Royal Society members and several Italian guests, commenting in detail on Rizzetti's critique.[69] With this, the discussion ended for virtually everyone.[70]

Although Newton believed that there was a Continental conspiracy against him, calling Rizzetti and other critics the 'friends of Mr Leibnitz', this seems not to have been the general point of view in England.[71] Desaguliers even ranked the German professor Richter with Newton. Moreover, Newton's paranoic belief does not provide a correct representation of the situation in the German lands: The stance adopted by Wolff, Richter and the *Acta eruditorum* towards Mariotte's and Rizzetti's criticisms was certainly not one of outright hostility towards Newton and his theory of colours.[72]

3

Theoretical traditions in physical optics, 1700–45

1. Development of the emission tradition

a. Newton's suggestions

The theoretical tradition in physical optics in which light is regarded as an emission of matter goes back at least as far as Greek Antiquity and has experienced a renaissance in modern times. Newton was a major representative of the emission tradition in seventeenth-century optics. In the Enlightenment period his ideas dominated this tradition, although he never held the position of an absolute ruler. Despite his unmistakably great importance for the eighteenth-century emission tradition, Newton had a curious status within it. He never unreservedly endorsed the emission hypothesis, in print at any rate. This is connected with his approach to methodology: He attempted to abstain from combining certainties with doubts. Consequently Newton's ideas on the emission of light are mainly encountered in the form of queries forming a supplement to his *Opticks*. His ideas remained suggestions that were unconnected with one another, provided no solution to some problems, and were occasionally inconsistent. In addition, they changed over time. For this reason we need a detailed study of the different versions of the 'queries' in the various editions of the *Opticks* if we are to obtain a clear and accurate account of Newton's ideas on the nature of light and of the later attempts to systematize his suggestions. The first edition (*Opticks*, 1704), the first Latin edition (*Optice*, 1706), and the second English edition of 1717, are the most important for our purposes.[1]

In the first edition of the *Opticks* (1704) Newton concluded the last part of the book, dealing with the inflection of light, with sixteen queries for consideration by others. In addition to inflection, these queries concerned colours, refraction, and reflection, Newton providing hypotheses for explaining all these phenomena.[2] In subsequent editions the queries were revised or increased in number. Thus, Newton enlarged the number of elements for a theory of light, and he even introduced nonoptical problems, such as the physical explanation of gravitation.

In 1704 Newton's conception of the nature of light is still full of vagueness. Consistently employing the question form, he attributes inflection, refraction, and reflection to an effect of material objects on light, without stating the nature of this effect or what light consists of. Conversely, the action of light on objects is supposed to cause parts of these objects to vibrate, thereby heating them. He talks of the emission of light by a warm 'fire' but says nothing about the nature of what is emitted.[3] He does not devote any attention to rival theories about light.

In the first Latin edition (1706) Newton added two queries containing a thorough discussion of his objections to other theories and hypotheses. The first query addresses the modification conception of colours, which, as we have seen, Newton had already rejected in 1672 on experimental grounds. He now rejects the general idea of explaining optical phenomena in terms of the modification of rays. Newton wonders, in a rhetorical vein, whether the phenomena of light are not dependent on 'original and unchangeable properties of the rays'.[4] He offers no examples here, but we may think of properties of the rays for the explanation of colours and double refraction, as we shall see shortly.[5]

In the second additional query of 1706 Newton aims his critique at medium hypotheses: 'Are not all hypotheses erroneous, in which light is supposed to consist in pression or motion, propagated through a fluid medium?'[6] Six arguments are provided to sustain Newton's criticism.[7] He distinguishes three kinds of medium conceptions according to the way in which light is propagated in the medium, that is, as a pressure, as an instantaneous (infinitely fast) motion, or as a motion with a finite speed. Four of the six arguments concern medium hypotheses in general, while the remaining two relate to specific medium conceptions. Newton's discussion of medium hypotheses is thus comprehensive and sophisticated, certainly by comparison with some other treatments from within the emission tradition.

Newton's first objection to a medium hypothesis is an extension of the previous query: Since all existing medium hypotheses explain optical phenomena by assuming modifications of the light rays, those hypotheses cannot be correct.[8] An obvious example, which appears later in the text, is double refraction in Iceland crystal (calcite). Huygens had explained double refraction by assuming that the pulses in the crystal were propagated in two different ways, that is, in spherical and spheroidal waves, resulting in a splitting of the light beam into two different ones. However, when he made a light beam pass through two crystals instead of only one, he observed, for certain orientations of the crystals, that the beam did not split again in the second crystal. Huygens himself concluded that the pulses in the ether suffered a modification in passing through the first crystal. However, he felt unable to identify the nature of this modification, which is hardly surprising considering that his theory recognised only longitudinal pulses. Newton levels his criticism against this aspect of the argument, believing he can prove that the path of light in the second crystal does not depend upon a modification of the rays caused by the first crystal, but upon innate and immutable properties of the rays.[9] Newton's explanation is that rays of light (which he appears to identify with particles of light) have four different sides, each of two pairs having the same properties. If one of the pairs points in a particular direction, the ray refracts in the normal way. If the other pair points in the same direction, extraordinary refraction occurs. The first crystal separates the two kinds of particles in a ray, producing the differing character of the two rays leaving the crystal. This idea enabled Newton to give an explanation in broad outline for the phenomena found in the case of two double-refracting crystals, whereas Huygens himself had confessed that he had not been successful in this.[10]

Newton's second and third objections are levelled against specific forms of the medium hypothesis. If light is supposed to be *pressure* in a medium, the heating of bodies by light cannot be explained, because a pressure cannot cause the particles in bodies to vibrate (in Newton's view heat consists of the vibration of these particles). The third objection is directed at the view that light consists of an *instantaneous* motion in a medium. In that case, Newton argues, an infinitely great force would be required to cause light to be propagated at an infinite velocity.[11] These two one-liners constitute the critique of the specific medium conceptions; the remainder of Newton's arguments concern the medium tradition in general.

The fourth argument could be designated the main objection, since Newton had already advanced it in an emphatic manner at an earlier date. In 1672, in response to an article by Hooke, he wrote:

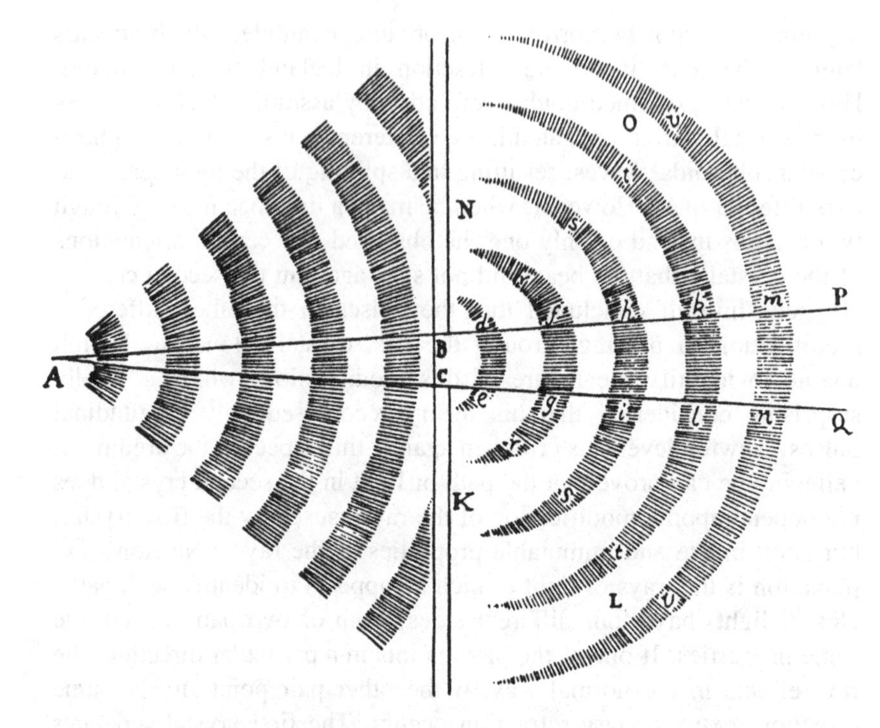

Figure 2. Newton on the transmission of a pressure or motion in a fluid medium (Newton, *Principles*, vol. 1, 368).

> . . . to me the fundamentall supposition it selfe seemes impossible; namely that the waves or vibrations of any fluid can like the rays of Light be propagated in streight lines, without a continuall & very extravagant spreading & bending every way into ye quiescent Medium where they are terminated by it. I am mistaken if there be not both *Experiment & Demonstration* to the contrary.[12]

Here Newton is claiming that both theory and experiment can demonstrate the nonviability of the medium hypotheses, yet at that time he provides no further information. He only published the theoretical part of his argument fifteen years later in the *Principia* of 1687 (see Figure 2). Newton claimed to be able to demonstrate that a pressure or motion propagated from a point *A*, through an opening *BC*, finds its way into the geometric 'shadows' *NBP* and *KCQ*.[13] He repeated this conclusion in a query in the *Optice* (1706) although without reference to the

Principia. At that time he produced several empirical indications that 'fluids' do actually behave in the way the theory expects them to do. Water waves encountering an obstacle gradually penetrate into the quiet water behind the impediment. Sound pulses or waves also penetrate into the 'shadow', although to a lesser extent than water waves. The sound of a bell rung or a canon fired behind a hill can be heard, even though the bell and canon themselves are out of sight. Moreover, sound is transmitted through crooked pipes as easily as through straight ones. However, unlike sound, light does not penetrate into the geometric shadow, as can be seen from solar eclipses and the disappearance of stars behind an intervening planet. The small degree of bending of light rays passing an object at a short distance is no exception to this rule; after all, the bending of rays occurs only when the beam is actually passing the object, and the beam follows a straight path again, once past the obstacle.[14] Therefore, we may add, light cannot be a pressure or movement in a fluid medium. Taking together these texts by Newton, we have the classic argument concerning rectilinear propagation in its most complete formulation. Though very interesting thoughts and experiments were developed in the century after Newton, his formulation of the argument remained the basis for the discussions (see, especially, Chapter 5, Section 3a).

The fifth objection concerns the explanatory mechanism of a new property of the light rays, the fits (*vices* in Latin). Newton believed that his experiments had demonstrated the existence of these fits just as surely as they had proved the heterogeneity of white light. He used the fits to describe the simultaneous occurrence of refraction and reflection (partial reflection) and the colours of thin layers, as in 'Newton's rings'. According to Newton, a ray of light at the interface of a medium is either in a 'fit of easy transmission', such that the ray passes on, or is in 'a fit of easy reflection', in which case reflection occurs. Newton concludes from the regular succession of the rings that in a ray of light the two kinds of fits are alternatingly and at equal intervals repeated. He provides an explicit hypothetical explanation of the fits by way of exception in the main text of his book, for 'those unable to accept something new or recently discovered, if they cannot immediately explain it by a hypothesis'. He posits that rays of light falling on a refracting or reflecting surface produce vibrations in the solid parts of the refracting or reflecting medium or substance that are propagated faster than a ray of light, drastically influencing its motion.

When a vibration 'conspires' with the motion of a light ray, the ray passes on, whereas if the vibration impedes the motion, the result is reflection.[15]

In his queries Newton suggests, as an argument against the medium theories, that these cannot explain the experimentally demonstrated fits. Here Newton's premise is that the fits can only be explained by a medium hypothesis if the existence of two interacting ethers is assumed, the vibrations in one forming the actual light while the periodic vibrations in the other take the light vibrations into alternating fits. Arguing against this, Newton points out that there is even no overriding reason for assuming the existence of only one ether, that the two ethers postulated would merge into one, and that the two kinds of vibrations would also involve the existence of two kinds of light.[16]

Furthermore, says Newton in his sixth and last argument, a great problem arises from filling the heavens with 'fluid' media, since from the regular movements of planets and comets it is clear that they encounter no noticeable resistance and that the universe therefore contains no 'sensible matter'. In the *Principia* Newton had stated that the universe is completely empty, 'with the possible exception of extremely tenuous vapours and travelling rays of light'.[17]

The last objection and particularly the fourth (concerning rectilinear propagation) produced a great response in the eighteenth century. The remaining four arguments did not attract much attention nor resurfaced in this period among the authors I have studied. Let us recall these objections. The first argument concerned the modifications of light, particularly in double refraction; the second argument dealt with the heating of substances, and the third with the force needed for instantaneous motion; the fifth related to the idea of two ethers as the explanation for the fits. There is a simple explanation for the disappearance of Newton's first objection, at all events where double refraction was concerned. In the eighteenth century double refraction was regarded as an isolated phenomenon, one of a great number of curiosities. There was little research on the problem, and textbooks dealt with the matter cursorily if at all. This also applies, although to a lesser extent, to the fifth objection. Partial reflection and Newton's rings were discussed but often only in a descriptive and qualitative way, without mentioning the fits or their hypothetical explanation.[18] The second and third objections lost their validity because the limited and specific variants of the medium tradition with which they were concerned ceased to exist after a while or had come to have only antiquarian value even in Newton's lifetime.

After dismissing the medium tradition in the manner described here, Newton expounds his emission concept, beginning – in the form of questions, naturally – with the main point: 'Are not rays of light small particles emitted by shining substances . . . ?'[19] Newton then summarises the advantages of the emission hypothesis. The rectilinear propagation of light follows directly from the basic conception of light particles being emitted and travelling rapidly. Light particles are also capable of possessing various immutable properties; Newton offers three examples. He explains the colour and refrangibility of light rays by assigning different sizes to the light particles. The smallest particles produce the colour violet, 'the weakest and darkest of the colours'. Small particles are easily diverted from their path; that explains why violet rays exhibit the greatest refraction.[20] The fits of easy transmission and easy reflection are explained by the assumption that the attractive powers of the light particles or 'some other force' stimulate vibrations that subsequently speed them up and slow them down at equal intervals. The different sides of the light particles – which Newton compares here with the poles of a magnet – are responsible for creating the particular phenomena observed in Iceland crystal.[21]

In refraction, reflection, and inflection the transparent objects act on the light particles, while these particles in their turn act on objects by heating them. Newton states that this 'action and reaction' strongly resembles an attractive force.[22] He introduces 'a repulsive force' only in the last and longest of the queries. The existence of a repulsive force 'seems to follow' from the reflection, inflection, and emission of light. He also considers this force to be responsible for the elasticity of air. Although he makes no mention of it here, he had already inferred Boyle's law (pressure times volume is constant) in the *Principia* by assuming that neighbouring particles of air repel each other with a force inversely proportional to the distance. He presented the repulsive force as a mathematical hypothesis, without wanting to commit himself to its physical reality.[23] In the *Optice* Newton goes beyond the cautious position adopted in the *Principia*; he states in the *Optice* that short-range attractive and repulsive forces exist, although the nature of their cause is not known, as in the case of the gravitational force that certainly exists but whose cause is still obscure. (Seven years later, in the well-known general scholium first included in the second edition of

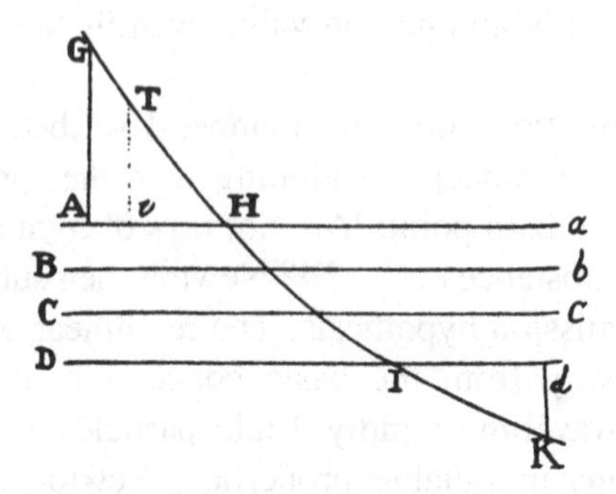

Figure 3. Newton's view of refraction (Newton, *Principles*, vol. 1, 228).

the *Principia*, he repeats this view on gravity, adding that he does not feign hypotheses on the cause of gravity.)[24] Since Newton only introduces the distinction between forces and their causes in his last query, the strong impression left by what goes before this, is that attractive and repulsive forces are inherent in matter and require no further explanation in physics.

Another result found in the *Principia*, the derivation of the law of refraction, undergoes the same change as the treatment of the elasticity of air (see Figure 3). In this work Newton had inferred the sine law for a hypothetical 'body' moving under the influence of an attractive force conceived as a mathematical and hypothetical entity. This force acts within a narrow zone *AadD* on both sides of the interface between two media. Newton also demonstrated that, where the angle of incidence is too large, the body is reflected internally, the angle of incidence being equal to the angle of reflection (see Figure 4). Up to that point Newton had not linked the hypothetical system of 'body' and area of force with the physical phenomenon of light. Only in a concluding scholium did he point out the mathematical resemblance to the refraction and reflection of light: his imaginary 'body' obeying the same laws as those governing a ray of light. Moreover, he claimed that in the inflection of light rays, then a recently discovered phenomenon, the rays of light bent as though they were affected by an attractive force. The rays passing most closely to the object were bent the farthest 'as if they were more attracted'. Because of this he thought that not only inflection but also the refraction of a light ray occurred gradually rather than at a point, for instance partly in the glass and partly in the air. He emphasised that he was only drawing an analogy between the propagation of light rays and the motion of bodies 'arguing absolutely nothing

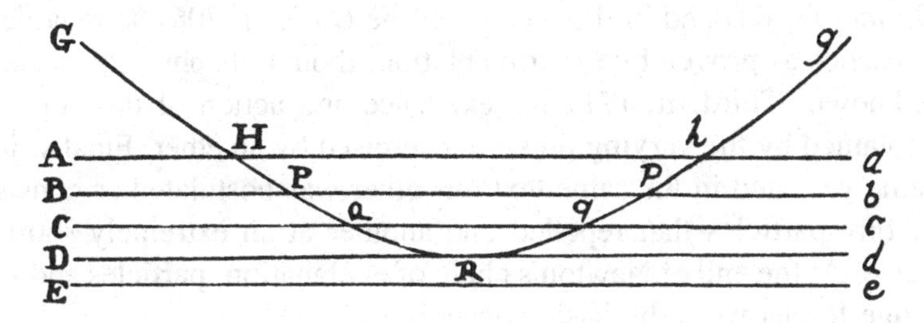

Figure 4. Newton's view of reflection (Newton, *Principles*, vol. 1, 229).

about the nature of the rays (whether they are bodies or not)'.[25] Within the new context of the *Optice* queries, where in this case he does refer to the *Principia*, the hypothetical bodies are changed into real light particles, and the mathematical area of force is changed into a real force of attraction.[26]

In the second English edition of the *Opticks*, published in 1717, Newton added a sentence to the end of the query stating his emission conception of light. He refers to new queries in which he elucidates what he means by 'vacuum' and 'attraction'. These queries turn out to be concerned with an ether that offers a physical explanation for the existence and action of the attractive and repulsive forces.[27]

The attraction when light is refracted from air to glass is traced back to a gradual rarification of the ether, which creates a pressure upon the light particles, resulting in refraction towards the normal. Inflection of light is interpreted in a similar way, by an ether density gradient, while the medium of the fits is identified with the ether.[28] Newton also considers that gravity can be explained by means of the ether pressure hypothesis. He attempts to repudiate the obvious counterargument, that the ether would hinder the course of the planets to a noticeable extent, which was the same objection he himself had raised in his sixth argument against the medium hypotheses. His ether is not a 'fluid' medium of the kind he considered earlier, but extremely 'rare', a term he clarifies when suggesting that the ether consists of extremely small particles, perhaps even smaller than light particles, that – 'like our Air' – repel each other at a very short distance. This repeats, on a smaller scale, his hypothesis on the nature of air.[29]

With his ether hypothesis of 1717 Newton provided the last link in a long chain of concepts to describe and explain refraction. First, in 1687 he described refraction mathematically in terms of a hypothetical attractive force operating at a distance. Second, in the main text of the

Opticks (1704) and in the queries of the *Optice* (1706) he regarded its existence as proven by experimentation, though its physical cause was unknown. Third, in 1717 the existence and action of this force was explained by the varying pressure exercised by an ether. Finally, in the same year and in the same text this ether was postulated as consisting of tiny particles that repelled one another at an extremely short distance. At the end of Newton's chain of explanation, particles and short-range forces were the basic concepts.

Newton's chain of explanation was open to a variety of interpretations, because its end-point could be located at different places, the choice of location being referred back to one of Newton's texts. First, if the person reading Newton adopted the viewpoint expressed in the *Principia*, the main text of the *Opticks* (1704), and the queries in the *Optice* (1706), the phenomenon of refraction would then be described in terms of a mathematical force of attraction, the cause of which (although the force definitely existed) remained unknown. Willem Jacob 's Gravesande, for example, adopted this position in his textbook. In a second interpretation one could (having devoted less careful attention to the queries of 1706) accept the actual existence of forces, at the same time believing that these were inherent in matter and that there was consequently no need to seek any deeper physical cause. Petrus van Musschenbroek, also an important writer of 'Newtonian' textbooks, opted for this interpretation.[30] The third and last option was (with reference to Newton's text of 1717) to trace refraction, or the refractive force, to the action of a subtle ether – its elasticity being due to short-range forces between the ether particles. In this interpretation the whole of Newton's chain of explanation was accepted, and refraction was ultimately reduced to the density of ether particles (its elasticity sometimes being ascribed to inherent microforces). This most complete account of refraction was the least accepted. Between 1700 and 1840 less than 10 per cent of the British and Irish champions of the emission hypothesis adopted a dynamic ether.[31] As we shall soon learn, this conception also found little response in the German lands.

Even if we bestow the name of Newtonian on all three conceptions of refraction discussed here, because they occur in Newton's work and were adopted by others, often with reference to Newton, then the post-Newton emission tradition was not exclusively 'Newtonian'. Thus, in 1717 Jean Jacques Dortous de Mairan used an emission hypothesis in which the concept of refractive force was totally lacking. For him the origins of refraction was a difference in the resistance of the two media. He also believed that space is entirely filled with a subtle ether; this idea is regarded by many historians as alien to 'Newtonian' thought, if not actually in conflict with it.[32] As an example of the 'non-Newtonian' variants within the emission tradition, in the next section I shall discuss de Mairan's conception. His version of the emission hypothesis, as well as other non-Newtonian versions, had far less influence than the 'Newtonian' mainstream, which I shall subsequently discuss. I shall examine two major representatives of this, 's Gravesande and van Musschenbroek, to demonstrate that even the mainstream did not have a uniform character.

b. De Mairan

In 1717 the French natural philosopher Dortous de Mairan published a treatise on luminescence[33] in which he introduced his subject by a lucid discussion of recent conceptions of the nature of light. At that time he favoured an emission hypothesis, having developed his ideas to approximately the same extent as Newton, from whom they were partly derived. Five years later de Mairan was to abstain from giving a public opinion on the nature of light, instead stressing the mathematical equivalence of the emission and medium hypotheses (see Section 3a).

In 1717 there is no hint of this neutral position. In contrast, de Mairan emphasises the sharp contrast between the two hypotheses on light:

> Tous les systemes modernes sur la Lumiere peuvent être reduits à deux. Car, ou les mouvemens du corps lumineux sont transmis jusqu'à l'oeil, seulement parce qu'ils se communiquent à la matiere qui est entre le corps lumineux & nous . . . ; ou l'agitation du corps lumineux produit en lui une émission, & un écoulement de corpuscules, qui viennent frapper l'organe de vûë. . . . Il n'y a pas de milieu dans cette alternative, ou le corps lumineux renvoye vers nous des particules de sa substance, ou il n'en renvoye pas; il faut necessairement que la Lumiere se répande de l'une de ces deux manieres, ou avoir recours aux qualitez occultes.[34]

A striking feature of his presentation of the problem is the self-confident way in which he accounts for the propagation of light, using the concepts of matter and motion and contrasting his account with the old explanations, which depended on occult qualities. However, something else deserves attention. Because of the exclusion of all hypotheses except the mechanistic, there were no obvious alternatives to the emission and medium hypothesis. Thus, two, and only two, possible hypotheses on light are considered. De Mairan seems to have been the first to give the either–or formulation of the problem that was to shape the discussion on the nature of light to a great extent over the next two centuries. Newton, for example, acknowledged the nonmechanistic hypotheses on light by using the distiction between hypotheses supposing modification of light and those that did not (above, Section 1a).

De Mairan's case illustrates one of the rhetorical consequences of the either–or problem formulation, that is, that the refutation of one conception (in this case, the medium hypothesis) is considered to have proved the validity of the other (here, the emission hypothesis).[35] De Mairan attempts to rebut the medium hypothesis by reference to the 'indisputable' fact that light is propagated in straight lines. In his view the medium hypothesis conflicts with this fact. The form in which he couches his objection differs from that employed by Newton:

> Si la propagation de la Lumiere se fait par les pressions des corps lumineux sur la matiere Etherée, je dis qu'il est impossible, qu'il y ait jamais de nuit ni d'ombres dans l'Univers.[36]

Newton had referred to rare solar eclipses, but de Mairan accentuates the objection by drawing attention to the frequent alternation of day and night and the universal fact that objects cast shadows.

De Mairan argues in favour of the claim that a motion or pressure in the medium is not propagated in a straight line by a reasoning different from Newton's, a feature that can be partly ascribed to the fact that he had Malebranche's conception of light in mind as his prototype for a medium hypothesis. Malebranche saw the propagation of light as the motion of periodic 'pressure vibrations'. To illustrate his hypothesis, he compared the luminiferous ether with a fluid placed in a spherical vessel and under great pressure. In his comparison the light source was represented by an oscillating piston.[37] De Mairan argues that Malebranche's conception conflicts with a known property of fluids:

> . . . en tous fluide renfermé dans un vaisseau dont il remplit exactement la capacité, la pression ou l'impulsion qui survient à quelques unes de ses parties, par l'action de quelque force que se soit, se communique à toutes les autres, & à toute la surface interieure du vaisseau, qui le contient.[38]

De Mairan applies his argument to the solar system, the pressure vibrations from the sun being transmitted throughout the whole of ether-filled space, which implies that the sun should be seen equally clearly over the whole earth. Further, the light in a room with a small opening would spread through the ubiquitous ether to every corner.[39] Even in a room without openings, with our eyes closed, we would perceive the light almost as clearly as we would in broad daylight, with open eyes. The reason for this, de Mairan says, is that, according to the medium hypothesis, a body is opaque not because it contains no ether but because the pores, which are full of ether, are bent rather than straight. Yet the propagation of a pressure within a fluid occurs just as easily in a bent tube as in a straight one, which would mean that no opaque bodies exist.[40]

In van Musschenbroek's work, which refers explicitly to de Mairan and Newton, we encounter de Mairan's poignant version of the main objection to the medium conceptions of light: the phenomenon of day and night, the shadows cast by objects, and the experiences with fluids in vessels.[41] The simplicity of the 'night argument' ensured that it became the classic objection to the medium tradition in textbooks and popular works. However, implicit in de Mairan's treatment of the subject was the assumption that a hydro*static* law or property can be applied to a hydro*dynamic* phenomenon. This assumption is closely related to the fact that de Mairan confines discussion to pressure hypotheses. Newton had not done so; consequently he treated the hydrostatic and hydrodynamic cases separately, particularly in the *Principia*, even though in both cases he reached the same conclusion. As we shall see, both 's Gravesande and van Musschenbroek considered only a pressure conception, thereby exchanging Newton's nuances for de Mairan's clarity and simplicity.[42]

De Mairan's emission conception also diverged from Newton's in its content. Without mentioning Newton or any of his early followers by name, de Mairan expresses his disagreement with several of the hypotheses often linked with an emission conception of light. De Mairan assured his readers that he rejected the idea of the vacuum, of atoms, an innate gravity, or mutual attraction and repulsion. He regarded these notions as fantasies, bringing discredit upon a conception of light that in itself was very probably correct. In his view the emission hypothesis amalgamated very well with 'le systeme du monde tel que les Cartesiens les plus modernes, & la pluspart des Astronomes le suposent'. Consequently he believed that all of space was filled with an ether that whirls around the sun and is responsible for the motion of the planets.

In that ether-filled space a luminous object gives out *another* subtle substance in a 'flux successif', just as a fountain produces streams of water.[43]

At this point de Mairan discusses two possible objections to his own emission hypothesis. The first concerns his ether and is therefore not applicable to most of the other ideas within the emission tradition. The objection engages the question of how the light particles can be propagated without being retarded or even stopped by the ether. De Mairan's reply is brief: The ether offers very little resistance; one of his arguments for this is based on a comparison with the motion of a cannonball through the air.[44] The second objection applies to every emission hypothesis, not only to de Mairan's: How can the sun give light for so many centuries and over such a huge area without diminishing in size or even being extinguished? De Mairan considers two possible explanations that may both apply. First, it is conceivable that light matter is extremely tenuous and that therefore the extent of decrease in the sun's size is imperceptible. Second, loss of solar material may be compensated by the return flow of light matter (from the stars, for example), and other matter or particles of the celestial ether may be changed into light particles in the sun.[45]

These, then, are the principle features of de Mairan's emission conception of light. The elaboration of his ideas and his explanation of luminescence, which is his ultimate goal, contain a few more characteristic traits. His hypothesis acquires a 'chemical' tinge when he identifies light matter with a very subtle and extremely mobile Sulphur, which is not the usual substance of this name, but 'le principe actif des Chimistes'. He has no hesitation in providing details of the light particles, which he believes to be spherical in shape, revolving round their axes, and variable in size. Below a certain minimum size they are no longer perceived as light, but the light particles are certainly much larger than particles of the celestial ether.[46] He believes the identification of light and Sulphur to be confirmed in his fairly comprehensive survey of luminescent substances that, he claims, usually contain a great deal of Sulphur. The process that sets the Sulphur in motion – the polishing of a diamond, or the rotting in wood, for instance – is the immediate cause of luminescence.[47]

De Mairan's explanation of reflection, refraction, and colours is scattered throughout the chapters on luminescence. Like Newton, whose name is still absent, he believes that reflection occurs somewhat before contact with a surface. He explains this by assuming the existence of an ethereal matter covering the tangible surface of the object and filling the pores and unevennesses like an extremely fine layer of varnish.[48] It is unlikely that he was inspired here by Newton's ether hypothesis, which was published in the same year as de Mairan's treatise, since de Mairan talks of a collision rather than of a gradual change of direction influenced by an ether density gradient.[49]

De Mairan traces refraction back to the differential resistance between two media. Without any further explication he posits that the ratio of the resistances is in inverse proportion to the ratio of the speeds. It is not clear whether de Mairan believes the speed of light to be greater or smaller in glass than in air, though he hinted at the latter. If this were the case, his view would differ from Newton's but he would be in agreement with several other emission theorists of the seventeenth century (see Table 2 in Section 3a). De Mairan suggests in a later article that the resistance in glass is less than that in air, the speed in glass consequently being greater than that in air, which was also Newton's assumption. In de Mairan's view the lesser resistance in glass does not conflict with the greater density of the glass, because the propagation of light takes place in the ether between the glass particles rather than in the particles themselves. This ether could be less dense in glass and therefore have less resistance.[50] Although both Newton and de Mairan use ethers, there are significant differences between them. Newton derives refraction from a difference of pressure in the ether, whereas de Mairan explains refraction by a change in resistance. The nature of the ether is also completely different: De Mairan assumes contact action, whereas in Newton's theory the ether particles repel each other. Finally, Newton links the simultaneous occurrence of refraction and reflection with the alternating ether fits, whereas de Mairan introduces both 'refracting' and 'reflecting' ethers.[51]

De Mairan believes that the dispersion of colours is caused by a disparity in the sizes of light particles, which results in a difference in velocities following refraction.[52] This belief accords with that held by Newton. Since we are dealing here with colours, it is no surprise that Newton's name is now heard for the first time. De Mairan is full of praise for the British man of science, about whose experiments he observes: '[L]es ingenieuses experiences dont il s'est servi . . . pourroient toutes seules immortaliser un nom moins celebre que le sien'. De Mairan provides an outline of Newton's theory of colours and of the *experimentum crucis*, with several variations as examples of Newton's experiments.[53] It appears that he tried to repeat the colour experiments, apparently with success; as far as is known he was the first in France to repeat Newton's colour experiments.[54]

As de Mairan's emission conception shows, it was possible to assume the emission of light particles without at the same time accepting the notion of the 'Newtonian' vacuum and forces operating at a distance. De Mairan's divergent position within the emission tradition provides us with a corrective to the idea that this tradition was entirely 'Newtonian' in the eighteenth century. The importance of this corrective is enhanced by the fact that de Mairan formulated his ideas on light, not only after Newton's time but also partially in imitation of the famous natural philosopher.

c. 's Gravesande and van Musschenbroek

The physics textbook written by 's Gravesande, published in 1720–1, is one of the most important of the 'Newtonian' textbooks, contributing greatly to the dissemination of Newton's ideas on the Continent, as well as in Britain. The work is also of great importance for the emission tradition in physical optics. Many people, especially those in Germany, became acquainted with 'Newton's' emission theory in the systematic form given to it by 's Gravesande.[55]

About a sixth of his textbook is devoted to optics, light and colours being dealt with in the context of fire (*ignis*). This context is significantly different from the one in Newton's *Opticks*, where the term 'fire' is almost lacking. 's Gravesande seems to have been influenced in discussing fire by the natural philosopher and chemist Boerhaave. 's Gravesande begins by observing that, although various properties of fire are known, many of its features continue to be shrouded in mystery and he prefers not to feign hypotheses on the subject, echoing Newton's 'hypotheses non fingo'. Nonetheless he adds an impressive array of the properties of fire: It penetrates everything with ease; it moves with great rapidity (something apparent from astronomers' observations); it combines with bodies, warms them, and causes them to expand. Fire is attracted by objects at a certain distance (here he draws attention to the inflection of light), and it is present in all bodies. Although 's Gravesande does not state explicitly that fire is a material substance, he implies as much in this list of properties.[56] He does state that light is emitted fire: 'An object that emits light, that is, [that] drives the fire along in straight lines, is called luminous'; 'When the fire enters our eyes in straight lines . . . it excites the mental image of light'.[57]

Like Newton, 's Gravesande traces inflection and refraction to the action of a force operating at small, but not infinitely small, distances.[58] In his discussion of reflection he deals first with the case of light passing from glass to air, where, he believes, reflection is caused by the attraction exerted by the glass. Reflection during the transfer in the opposite direction, from air to glass, makes it clear that light does not collide with the particles in the glass, but is reflected at some distance. He claims that 'light is repelled at a certain distance from the objects, in the same way as the refracting force operates at a certain distance from the objects'. He mentions a 'reflecting force' but not the term 'repulsive force'. It is unclear where the spheres of influence of the refracting and reflecting forces are situated in the border area between two media.[59]

Up to this point 's Gravesande adheres to the suggestions made in Newton's queries, but he does not adopt the remaining 'hints'. Even more strange, we find no trace of the fits of easy transmission and reflection discussed in the main text of the *Opticks*, whereas 's Gravesande does give a treatment of the related subjects of partial reflection and the colours in thin layers.[60] Neither does 's Gravesande adopt Newton's suggestions concerning the sizes of light particles and their differently natured sides or 'poles'. This is very probably because, in accordance with his strict rules of methodology, he does not want to express detailed opinions on the material nature of fire and thus of light. Nevertheless, the subject of double refraction, certainly suited to a descriptive treatment, is not discussed at all. This subject 's Gravesande may have considered too difficult for treatment in a textbook. Finally, we find no trace of Newton's suggestion that the cause of the optical forces and the fits must be sought in the action of an extremely tenuous and elastic ether.[61]

's Gravesande's emission theory is a purified version of Newton's hypothetical ideas on the nature of light and colours. The basic concepts are emission and forces; he avoids discussion of very detailed propositions on light, and he does not introduce an ether. As we shall see in due course, other authors of textbooks had less difficulty in spelling out the various material properties of light particles, even if they did not go as far as Newton and were, like 's Gravesande, not prepared to introduce an ethereal medium.

It is striking that in the 1720–1 edition 's Gravesande never discusses other conceptions of light, such as the medium hypotheses. The situation changes in the third edition of his textbook, published in 1742, although discussion of rival theories is less comprehensive than Newton's.

's Gravesande advances three objections to medium hypotheses. Remarkably enough the argument that they cannot explain the propagation of light in straight lines has to be found between the lines of what is written on rectilinear propagation: 'If light passes through an opening it maintains its direction and is not dispersed to the sides, as has been said about waves'.[62] 's Gravesande does not explicitly attack the medium hypotheses on light, but his criticism is implied by the cited phrase. He goes much deeper into two other objections, neither of which is found in Newton's work.

These arguments are raised when 's Gravesande treats the nature of light. He perceives two questions requiring answers: first, the question of whether the motion of light ought to be conceived as a pressure or as a real motion, and second, the question of whether the propagation of light is instantaneous or gradual.[63] He deals only briefly with the first. Those who accept the idea of a pressure, he says, posit extremely small globules spreading throughout space, touching each other. However, a globule can be pushed in various directions at a particular moment. This means, among other things, that opposing pressures neutralise each other, yet experience teaches us that countless rays can pass through an extremely small opening without any mutual disturbance.[64] Pressure theories of light are thus rejected, and at the same time, 's Gravesande believes, the second question is answered. If light is not the transmission of a pressure, then it is an actual motion, and therefore the velocity of light is finite, a fact that is confirmed by the measurements of the speed of light made by Ole Römer and James Bradley.[65]

It is striking that, in his treatment, 's Gravesande only adopts one of Newton's arguments, that concerning rectilinear propagation – albeit implicitly rather than explicitly – and thereafter makes use of two different arguments. Even more surprisingly, out of all the available medium conceptions, he discusses only the pressure hypothesis, just as if Huygens had never existed.[66] We cannot reproach 's Gravesande with failing to do sufficient justice to the medium tradition. As we shall see in the next section, in 's Gravesande's time Huygens' theory, even in the medium tradition, had been forgotten or, if it was considered at all, was at best controversial. Descartes' view of light was generally conceived as a pressure hypothesis in which it was supposed that the propagation of light was instantaneous. Finally, in his presentation of the hypothesis of pressure vibrations, Malebranche was unclear about the velocity of light. For a textbook, 's Gravesande's exposition is certainly not without its nuances.

There were no great changes in the contents of the theory of light between the first edition and the third of 1742. In the latter 's Gravesande argues with greater clarity that the true nature of fire is unknown, but that fire is present wherever we observe heat and light. 'Fire' is replaced by the term 'light' in the formulation of the luminosity of objects: '[L]ight is emitted in straight lines by objects'. It is apparent that this is a purely terminological alteration from the fact that, for instance, he says that the inflection of light rays clearly shows that fire is attracted by objects.[67] As in the first edition, he omits Newton's fits and the latter's hints about the variable size of light particles, their 'poles', and the luminiferous ether.

There is greater clarity in his conception of forces. Here he writes explicitly about the 'repulsion' of light, and like Newton, he conceives a link with short-range forces between particles. The attraction and repulsion in inflection, refraction, and reflection are subject to the same laws as the short-range forces operating between particles. Attraction is greatest in contact, after which it diminishes until it turns into repulsion at a small but finite distance.[68] He recognises that the causes of the attraction and repulsion of light, which are realities, are hidden from us. Quoting from Newton's last query in the *Opticks* he posits that the forces, while not themselves of an occult kind, have occult causes.[69] In this 's Gravesande was more cautious than most 'Newtonian' natural philosophers, who regarded the forces as explanatory concepts. However, the subtlety of his attitude was not always respected, and on this issue he was often identified with the other 'Newtonians'. In this respect he fared no better than Newton himself.

To show that variations arose even within the mainstream of the emission tradition, we shall now examine the work of a second important author of 'Newtonian' textbooks, Petrus van Musschenbroek. He has a large number of volumes to his credit, in some subjects displaying considerable diversity.[70] Yet where optics is concerned the various works and editions do not show a great deal of difference from one another, with a single exception, as we shall see. The second edition of the *Elementa physicae*, appearing in 1741, is the most suitable work for an analysis of the content of van Musschenbroek's optics, partly because this book was translated into German in 1747 and was extensively cited by Germans in the second half of the eighteenth century.[71] As far as the first half of the century is concerned, 's Gravesande's books appear to have been of greater importance than van Musschenbroek's, although references were also made to the latter's first textbook, the *Epitome elementorum physico-mathematicorum* of 1726, and to the first edition of the *Elementa* appearing in 1734.[72]

The first thing to notice when comparing van Musschenbroek's treatment of light and fire with that of 's Gravesande is the difference in style. Van Musschenbroek is more widely ranging, poses more questions, and devotes greater attention to alternative theories of light. He is also less cautious about the nature of fire and light; stylistic differences thus merge into a difference in content.

Basing his view on that of Boerhaave, van Musschenbroek posits that fire is a substance consisting of extremely tenuous parts, each with a smooth surface. The particles of fire are exceedingly mobile; they adhere to bodies; and they have weight.[73] Light travels at great speed and in straight lines. It is a very 'fluid' substance, meaning that the particles cohere barely at all. With the help of a calculation he tries to demonstrate that light is a substance of such subtlety that it offers almost no resistance to the planets.[74] Van Musschenbroek assumes at these places a close relationship between light and fire, yet only later does he address the question of whether, and to what extent, light and fire differ from each other. He considers it probable that they are the same thing in both qualitative and quantitative respects.[75] He believes that the luminescence of the Bologna stone[76] originates in the 'drinking' of light by the stone, which subsequently emits the light again. He is strongly influenced by the experiments performed by Ehrenfried Walther von Tschirnhaus in which asbestos – incombustible in any earthly fire – had been changed into glass with the aid of burning glasses and burning mirrors. He accounts for the 'unbelievably strong effects' by the assumption that the action of converging rays is reinforced to a far greater extent than could be expected merely by adding together all the separate effects.[77] For van Musschenbroek fire is a very special substance because it has properties not found in other substances. His argument is that no other known substance is as subtle as fire or divides itself so evenly in the spaces between bodies. Moreover, experiment – he appeals to Boerhaave here – shows that other substances do not turn into fire.[78]

Van Musschenbroek's more 'speculative chemical' streak is apparent: He postulates kinds of matter that are qualitatively different. Later in the eighteenth century this conception of the nature of matter was almost universally accepted, with 'imponderables' such as heat, light, electricity, and magnetism: substances of little or no weight, distinguishable from each other in their physical as well as chemical properties. 's Gravesande's conception of light and fire is better suited to a theory of matter in which only one kind of matter is distinguished and in which chemical effects have their origins in quantitative differences of a qualitatively uniform matter.

Where the existence of forces of attraction and repulsion are concerned, van Musschenbroek's theory accords in broad outline with that of 's Gravesande.[79] Nonetheless, he adopts an entirely different attitude to the status of these forces. Van Musschenbroek believes that God has put them into bodies and that they were therefore properties of matter that need no further explanation.[80]

Like 's Gravesande, van Musschenbroek omits double refraction and consequently the 'polar' hypothesis. Van Musschenbroek diverges again from the views 's Gravesande expressed, in his discussion of colours and Newton's rings. His explanation for the diversity of colours is given in the spirit of Newton's suggestions: He believes that either the volume or the density of the light particles varies, although he leaves the choice open.[81] As we have seen, Newton opted for the first of these.

Van Musschenbroek gives a descriptive and qualitative account of Newton's rings, mentioning only that some rays of light pass through objects, whereas others are reflected, depending upon the thickness of the layer of air between two spherical lenses pressed together.[82] No reference is made to the periodic fits, the existence of which Newton claimed had been proved experimentally, and there is also no discussion of Newton's hypothesis that the waves in an ethereal medium are responsible for the fits. This treatment resembles that provided by 's Gravesande, but van Musschenbroek had earlier adopted a different view. In the first edition of the *Elementa*, appearing in 1734, he asks in a rhetorical vein whether a force (*vis*) exists that emanates from the surface of a substance, behaving like a wave; in the 'troughs' of the waves light is allowed to pass, while light is reflected at the 'peaks' of the waves (see Figure 5).[83] The undulating force was removed from the 1741 edition and did not return in his subsequent textbooks. This example illustrates how one of Newton's experimental discoveries (the

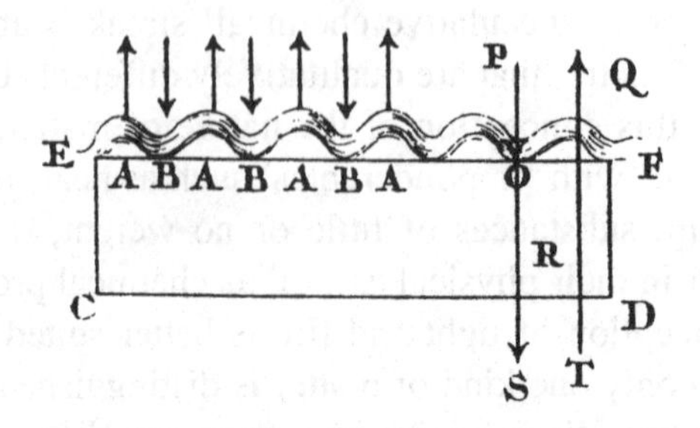

Figure 5. Undulating force in a refracting medium, after van Musschenbroek (van Musschenbroek, *Elementa* [1734], Table VII, Fig. 3).

fits) had fallen in disrepute or became progessively more unknown in the eighteenth century.[84] Even someone like van Musschenbroek, who had originally accepted the existence of periodic fits in a certain sense, no longer referred to them.

Like 's Gravesande, van Musschenbroek restricts general discussion of other hypotheses of light to a rebuttal of Descartes' pressure hypothesis, which represents 'the' medium hypotheses. Van Musschenbroek regarded the infinite speed of light, which he considered to have been proved by Römer's observations, as fatal to any pressure hypothesis. Further, according to that hypothesis there would be no shadows: When pressure is applied to one side of the fluid in a vessel, the pressure is transmitted in all directions; consequently, according to a pressure hypothesis of light, the sun would always be visible, whether above or below the horizon, but this conflicts with experience. The formulation of the disagreement suggests de Mairan's influence, although van Musschenbroek also seems to have based his argument on Newton's work.[85] With this the general discussion of other hypotheses on light is finished, although van Musschenbroek returns to the ideas propounded by Descartes and others, naturally with the intention of refuting them, for instance, in his treatment of refraction and colour.[86] We can also hear the echo of an objection to the emission hypothesis in van Musschenbroek's repeated assurances that light is extremely subtle and therefore does not hinder planetary motions.[87] However, he gives no explicit account of the potential disturbance of planetary motion as an argument against his own theory.

In his book *Beginsels der natuurkunde* (Principles of physics), of 1739, van Musschenbroek discusses several open questions arising from his theory, including the problem of explaining the emission of light. He weighs a number of possibilities, only to reject them all in the end. He concludes that the cause of the emission of light 'appears to be one of the things least known'. However, he is much more decisive in the *Elementa* of 1741. The 'questions' are not absent from this work, but they are presented in a shortened form, and they scarcely, if at all, formulate problems for the emission theory; they are plainly questions for further research.[88]

2. Development of the medium tradition

a. Descartes and Huygens

Descartes occupied a paradoxical position in the medium tradition. Although his name was frequently cited in the eighteenth century as a leading proponent of a medium theory, his conception of light diverged from the eighteenth-century medium theories in almost all its features. Only the underlying idea, the active role of the medium in the propagation of light, was the same. The fact that Descartes was cited as a representative of the medium tradition was derived from his status as one of the most prominent mechanistic philosophers. Anyone defending a particular mechanistic medium hypothesis felt himself to be following in Descartes' footsteps, even when that specific hypothesis showed very little resemblance to Descartes' actual contribution to the medium tradition.

In his main work, the *Principia philosophiae*, of 1644, Descartes' ideas on light form part of his cosmology and conception of matter. He believes the whole of space to be filled with matter, of which there are three classes, or 'elements', distinguishable from one another by their size and state of motion. The particles of the first element (the 'subtle matter') are the smallest and most mobile, and they are the material of which the sun and stars are composed. The second element consists of spherical particles whirling around the sun, which carry the planets along with them in their motion. The space between the spherical particles is filled with subtle matter. Taken together, the particles of the first and second elements are called the 'ethereal matter'. Terrestrial matter is composed of the largest and least mobile particles, that is, the third element. Descartes defines sunlight as the centrifugal 'tendency towards motion' (*conatus*) of the celestial matter in the solar vortex. This tendency is propagated instantaneously and in a strictly rectilinear fashion, which seems to solve the problem of rectilinear propagation of light.[89] Alas, the centrifugal tendency explains only the existence of rays parallel to the ecliptic. Moreover, the radiation is present without any contribution from the light sources; the presence of the vortex is sufficient.

In order to explain rays of light travelling in all directions and to allow the light source a role, Descartes assumes that an exchange of particles from the first element takes place between the vortices; they move out at the ecliptic and in at the poles. The particles entering at the poles make their way towards the centre of the sun, creating a pressure in all directions that is propagated instantaneously through the celestial matter. The rectilinearity of the pressure propagation cannot be deduced theoretically. It would appear that Descartes added this pressure (on the celestial matter from the first element) to his conception of light only at a late stage, yet this second idea was generally regarded as his only basic assumption. The centrifugal conatus was given no role by either supporters or opponents, and Descartes' conception was interpreted as a pressure hypothesis.[90]

It is striking that in his *Principia philosophiae* (1644) Descartes did not deduce the laws for refraction and reflection,[91] which had constituted an important part of the *Dioptrique*, of 1637. In the *Dioptrique* the discussion of reflection and refraction was not based explicitly on Descartes' basic idea that light was a tendency among the particles of celestial matter to move or to create a pressure. At that time he did not confront the question of the nature of light directly, but only used analogies or 'comparisons' (*comparaisons*). Descartes' particular application of these comparisons resulted in many misunderstandings in the seventeenth century and after.

In the *Dioptrique* he offered various comparisons that were apparently inconsistent. However, each comparison was meant to illustrate one particular property of light rather than to offer a comprehensive physical picture. This strategy enabled Descartes to satisfy the requirements of consistency with greater ease. One example is the relationship between his first and third comparisons. In the first – a blind man tapping with his stick to 'feel' his environment – Descartes introduced the instantaneous tendency to move as the first characteristic of light. In the third comparison, used for instance in his treatment of reflection, refraction and colours, he compares a ray of light to the path of a thrown ball or stone. Descartes says that both comparisons apply because, in a change of medium, the tendency to move and actual motion obey the same laws.[92] Elsewhere he explains that this does not mean that a tendency to move and actual motion have the same speed, which would imply that light is not propagated instantaneously; in the third comparison he is comparing the 'forces' and the 'paths', rather than the velocities.[93] In Descartes' elaboration of this comparison it is obvious that the velocity of the spherical particle is greater in glass than in air. This speed ratio and the comparison used fit exceptionally well into Newton's emission conception of light. Consequently it is probable that Newton's mathematical deduction of the law of refraction in the *Principia* and his physical model of refraction in the *Optice* are transformations of Descartes' comparison.[94]

Descartes' idea, cited in the previous chapter, that colours originate in the speed at which spherical particles rotate, is an illustrative comparison to make a particular point. First, he does not commit himself in the *Dioptrique* to the view that light consists in the action of a subtle medium, just as he does not adhere to the view that a ray of light 'really' is a blind man's stick. Second, once again, what matters to him is not an actual rotation, but the *tendency* to rotate. Although he makes this plain in one place in the *Dioptrique*, he gives no emphasis to the point, and in a passage of *Les meteores* he gives the impression that there is a real rotatory motion.[95] This impression is reinforced because in the latter context he does talk of subtle matter that passes on the action of light. This allowed the belief to take root that the spherical particles of the second element actually rotate. Supporters as well as opponents generally conceived Descartes' rotating spherical particles as physical reality, rather than as imaginary elements of an illustrative comparison.[96]

Descartes' theory of light cannot be considered identical with that of Huygens, published in the *Traité de la lumiere* in 1690. In Descartes' work we have found a 'pressure hypothesis' involving instantaneous transmission, observations on rays, not on pulses or waves, and deductions of reflection and refraction based on an analogy with the movement of a ball. In contrast, Huygens assumed a finite velocity of propagation; he used rays as well as pulses, and he based his deductions of reflection and refraction upon the behaviour of these pulses. Huygens' theory is regarded as part of the seventeenth-century 'kinematic tradition', which began with Thomas Hobbes; Hooke and Pardies carried it further, and it reached its peak with Huygens. The tradition is called kinematic because the description it provides of the propagation of pulses or waves is independent of the mechanical origins of motion and the possible periodicity of light. The central kine-

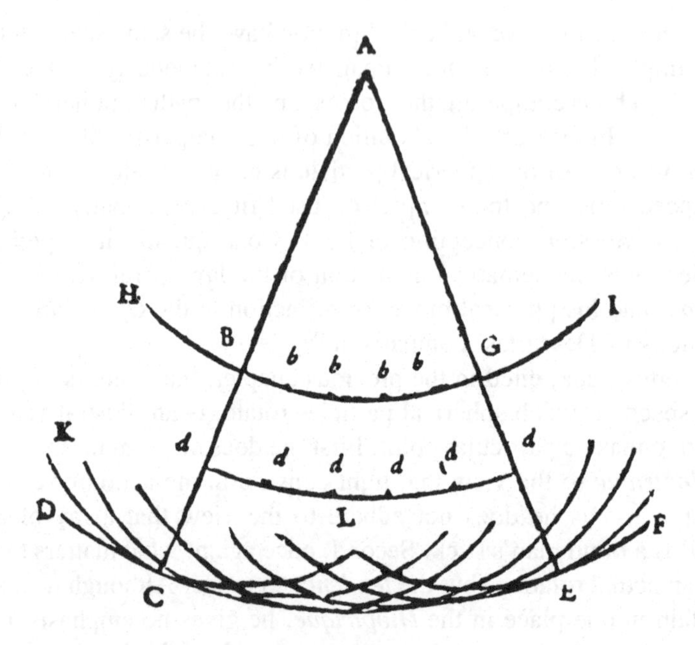

Figure 6. Huygens' principle (Huygens, *Traité*, 17).

matic concept is the 'wave front', or – more precise but not often used – a 'pulse front'.[97]

Huygens' medium, the light ether, consists of extremely small particles, not necessarily spherical, very hard and in contact with each other. The pulses in the ether are propagated at a great, but not infinite, velocity. Huygens overtly dismisses the possibility that the pulses follow on another at periodic intervals.[98] He subsequently introduces the 'pulse front' concept and 'Huygens' principle' (see Figure 6). Secondary spherical pulses travel from each point of the pulse front *HBGI*, the pulse from point *B* meeting the pulse front *DCEF* at *C*. The other pulses with their centres at the points labelled *b* also make a contribution to this pulse front. The individual pulses are 'infinitely' weaker than the pulse front they form collectively.[99]

He uses only the concepts introduced here in his discussion of the rectilinear propagation of light. If *BG* is an opening in an opaque object *HI*, the rays *BC* and *GE* then form the boundaries of the light cone because the secondary pulses outside *BCEG* do not concur to form a pulse front and are thus too weak to produce light.[100] As we shall see, eighteenth-century observers were not unaware of the fact that this explanation of rectilinear propagation of light was unsatisfactory. This is less immediately obvious for those in the twentieth century, since schoolbooks from our century often foster the impression that Huygens' principle is sufficient to explain rectilinear propagation, while failing to mention that Fresnel's principle of interference is also needed for this.[101]

Huygens' principle justifies its existence in its explanation of double refraction. The laws of reflection and refraction could still be inferred from the principle that pulse front and ray of light are perpendicular to each other, Huygens' principle being its equivalent in this derivation. However, double refraction can be explained only by means of Huygens' principle, and its application shows clearly that the extraordinary ray is no longer perpendicular to the pulse front. Because of this, Huygens called double refraction and its explanation his *experimentum crucis*.[102]

The historian of science A. E. Shapiro maintains that double refraction was decisive for the acceptance of Huygens' theory: Researchers who approved the theory did so on the basis of the explanation it provided for double refraction.[103] However, Shapiro lays insufficient stress on the fact that those who rejected the theory, even those from within the medium tradition, generally found the account of rectilinear propagation inadequate.[104] Other factors also played a role in this rejection: the restricted scope of the theory, which did not deal with the subject of diffraction, gave no explanation for the extraordinary phenomena observed in the case of two double-refracting crystals (polarization) and – most significant – provided no treatment of colour phenomena. Furthermore, in the eighteenth century interest was directed more towards dynamics than towards kinetics. Finally, no attention was devoted to double refraction and thus to the strongest point in Huygens' theory. The balance was soon tilted towards rejection of the theory, or at best simply a lack of interest. From about 1700 the *Traité* was almost completely ignored even in research reports from within the medium tradition.[105] However, Huygens' theory of light had a favourable reception in several German textbooks dating from the first half of the eighteenth century (see Chapter 5, Section 1a).

Huygens devotes barely any attention in his *Traité* to other theories of light. He refers to the phenomena observed in the case of burning glasses, amongst other examples, in defense of the idea that light is a 'motion'. This remark is obviously directed at ideas of a nonmechanistic kind. Having dealt with this obligation in the briefest of ways,

Huygens counters the mechanistic conceptions that differed from his own, and he confines his attention to a short discussion of the emission hypothesis, bringing out the fact that it conflicts with the immensely high velocity of light and especially with the unimpeded crossing of light rays. In a letter written in 1694 the same two arguments appear, but there they are linked. In order to avoid the collision of emitted light particles when rays cross each other, any emission conception has to include the assumption that light particles follow each other at a very great distance. In that case, asks Huygens, how can one explain the ultrarapid propagation of light? Here he obviously believes that an extremely fast propagation of individual particles is – a priori – unlikely and that only a 'chain collision' of particles not too far apart can provide a sufficiently great speed. In the same letter Huygens adds a third argument against an emission hypothesis. His correspondent Leibniz had raised the objection that he found difficulty in deducing the laws of refraction from light particles alone. Huygens agrees with him, adding that it seems to him that refraction in Iceland crystal – 'qui me sert d'*Experimentum Crucis*' – presents even greater problems.[106]

The 1690 edition of the *Traité de la lumiere* includes a *Discours de la cause de la pesanteur* in which Huygens responds to two of Newton's arguments against the medium tradition. The first objection is that the planets would encounter resistance from an ether unless this were extremely rarified. Were Newton's argument correct, the bottom would have fallen out of Huygens' theories on both gravity and light. In his reply Huygens concedes that the existence of a tenuous ether in the form of widely distributed particles separated by a great deal of space would be unable to explain gravity or light. He thinks, however, that there is a second way in which a fluid can be permeable to light, or present no resistance to it. Although the particles in a fluid touch each other, their great mobility ensures that they offer little resistance. Huygens gives the example of water, since like Descartes he believes that water offers less resistance than a solid substance, for the reason mentioned. Ether particles possess an even greater mobility than particles of water, and therefore present almost no resistance to the motion of the planets. Furthermore, there is no justification for Newton's objection that the weight of the ether obstructs motion, since it is the ether itself that causes the weight of bodies, and it is not heavy in itself.[107]

Newton's second objection discussed in Huygens' *Discours* is the main argument against a medium theory, namely, that this kind of theory cannot explain the rectilinear propagation of light.[108] Huygens does not dispute Newton's proposition that a motion within a fluid medium deviates from a straight line. Yet in his view the action of the lateral waves or pulses becomes progressively weaker in the sequence of water–air–ether. He agrees with Newton on the subject of water waves: They are perceptible in the 'shadow', although Huygens maintains that the lateral waves are weaker than those travelling in the main direction. He repeats that lateral light pulses have no effect on the eyes. Strangely, the further discussion concerns only sound. Newton's empirical support for his ideas on wave motion taken from the phenomenon of sound propagation, as this appears in the *Principia* of 1687, is challenged by Huygens in some detail. Newton posits, directly after his proof that pulses propagating in an elastic medium penetrate the 'shadow':

> We perceive this in sounds which are heard, either when there is a house interposed, or, when they are admitted into a room by the window and spread into all parts of the room and are heard in every corner, not being reflected from the opposite walls but being transmitted directly from the window.[109]

Huygens uses two different counterarguments in his reply. First, he disputes Newton's conclusion that sound propagates nonrectilinearly on the grounds that sound can in both cases reach the listener by means of reflection and can nevertheless come from the immediate vicinity of the sound source or the opening. Where the room is concerned, sound can reach our ears from a place close to the window after a series of reflections occurring as though in a single moment. Second, he introduces a new phenomenon, the echo, into the discussion. If Newton were correct, echoes would not meet the equality of the angles of incidence and reflection so accurately. For

> . . . quand on est placé en un lieu, d'où il ne peut point tomber de perpendiculaire sur le plan reflechissant d'un mur un peu eloigné, on n'entend point repondre l'Echo au bruit qu'on fait en ce lieu, comme je l'ay experimenté tres souvent.[110]

Huygens provides no further information on his observations. Even in a manuscript on the measurement of the speed of sound by way of an echo, there is no specification of the experiments under discussion here.[111]

Huygens' reaction resulted in minor changes in Newton's formulation of his main objection to the medium tradition. In the second and third editions of the *Principia* the term 'house' is replaced by 'mountain'. In an addition to the passage on the room, he states that sound comes from the direction of the window 'as far as our sense can judge'. Newton takes his biggest step in Huygens' direction in the *Optice* in which, as we have already seen, he posited that sound waves do actually bend in the 'shadow', although not as much as water waves.[112] Here Newton accepts that there is a difference in degree between the ways in which the waves in water, air, and light ether behave. Nevertheless, his objection to the medium tradition remains essentially intact.

Huygens' answer to Newton continued to exert influence through the eighteenth century, although without mentioning the former's name, as a rule (see Chapter 5, Section 3a). Huygens' position within the medium tradition was as ambiguous as that of Descartes. Although from the nineteenth century to the present Huygens has been seen, and lauded, as one of the founders of 'the' wave theory of light, we encounter his name only sporadically in the eighteenth century. Where Huygens was mentioned it was usually in the same way as that in which Descartes was, that is, they were both seen as authorities with the same basic ideas. Huygens' principle, the idea that was special to Huygens' theory, was barely accepted, and as early as the beginning of the eighteenth century it was dismissed entirely.

b. Malebranche and Johann II Bernoulli

Nicolas Malebranche's ideas on light and colour, which changed considerably over a forty-year period, reflect the rapid development of the medium tradition at the end of the seventeenth century and the beginning of the eighteenth. Malebranche first adopted Descartes' pressure hypothesis and the modification theory of colours (1675), subsequently introducing the concept of the frequency of ether vibrations, guided by the analogy between sound and light (1688, 1699). He finally produced an interpretation and explanation of Newton's theory of colours from within the framework of a medium hypothesis (1712).[113] The early ideas were expounded within a general philosophical context and were consequently treated only in rough outline. The idea of linking colour and frequency was also not initially worked out in detail (1688),[114] but in 1699 he did provide a full exposition, in a lecture given to the Paris academy, that was entirely devoted to the subject of light and colours.[115]

In 1699 Malebranche maintains an open mind on the question of whether the speed of light was very high or actually infinite, although in 1712 he was to incline to the idea of infinite velocity.[116] He also believes that the globules of light ether could not be hard, as Descartes had posited, for, Malebranche writes, if they were indeed hard then one globule at the point where two rays of light cross would have to turn on two axes at once in order to pass on the 'colour rotations'. Malebranche believes the ether particles to consist of vortices (not necessarily spherical in shape) of an even more subtle, extremely fluid and mobile matter.[117] Because of the centrifugal force of the microvortices, the ether is a highly elastic medium exerting a constant pressure. However, just as the constant air pressure does not result in any perception of sound, our eyes do not perceive the constant pressure of the ether. Only changes in the equilibrium state of the ether – pressure vibrations (*vibrations de pression*) – are experienced as light. The cause of pressure vibrations in the ether is the extremely rapid movement of the tiny particles in a light source. The 'stronger' the vibrations, the clearer the perception of the light source. The 'speed' of the vibrations (i.e., the frequency) determines colour.[118] In 1699 Malebranche was a supporter of the modification theory of colours, white being a colour like any other in his view. White had the greatest frequency, followed by yellow, red, and blue; black had frequency zero. Although the frequency ratios of musical consonances were known, Malebranche considered it impossible to determine the relative number of vibrations in the case of light.[119] In this way Malebranche expressed in 1699 the idea of linking colour to the frequency of ether vibrations (thus reminding us of Newton's suggestion to Hooke).[120]

Nevertheless, as a systematic and developed theory of light and colour, Malebranche's ideas leave much to be desired. There is no discussion of reflection and refraction; the rectilinear propagation of light is not recognised as a problem; discussion of the transmission of pressure vibrations is superficial and nonmathematical. Only the most basic principles of a theory of colours are given; no explanation is provided for permanent colours or colour dispersion.

More than ten years later, in 1712, Malebranche published an amended and extended version of his ideas. Newton's *Optice* of 1706 was the source of the alterations and supplementary material, especially in the treatment of colours.[121] In a letter dating from 1707 Malebranche says about this book:

> Quoique Mr. Newton ne soit point physicien, son livre est tres curieux et tres utile à ceux qui ont de bons principes de physique, il est d'ailleurs excellent geometre. Tout ce que je pense des proprietez de la lumiere s'ajuste à toutes ses experiences.[122]

Despite Malebranche's phraseology, which sounds somewhat patronizing to the modern ear, he understood Newton's aspirations better than most. Newton had not wanted to provide physical hypotheses in the *Optice*, at any rate in the main text, only properties and laws deduced from observations and experiments. For Malebranche the framing of physical hypotheses was the sole task of the 'physicien' (natural philosopher). The performance of experiments, without a hypothesis on the nature of light to provide a physical explanation of the phenomena, was – strictly speaking – not part of his task. Thus, Malebranche was not contradicting himself when he stated that Newton was no 'physicist' at all, while at the same time singing his praises as an experimenter and mathematician (also occupied with the mathematical aspects of physics and astronomy).[123] Malebranche consequently adopted Newton's empirical conclusions, explaining them in the light of his own conception of the nature of light; this conception was, of course, based upon 'good physical principles'.

It was a fairly simple matter to fit Newton's empirical results into Malebranche's theoretical ideas. In 1712 Malebranche dropped the modification theory of colours and adopted Newton's idea of seven simple or homogeneous colours, which he distinguished according to their frequency. Newton had divided the borders of the seven colour bands in the solar spectrum into a kind of musical octave, and Malebranche now adopted this idea in a somewhat altered form by dividing the frequencies harmonically. He implicitly assumed that they fit into one octave. Thus, in his own view he was providing a solution to the problem that had previously seemed to him insoluble, that is, the way of determining the ratio of the frequencies. He also thought he could determine the colour having the greatest frequency. His argument was obscure, appearing to trace a connection between the 'strongest' colour, red, and the circumstance that vibrations starting less frequently have a greater amplitude (a sign of 'strength'?). Whatever the truth of the matter, he concluded that red has the lowest frequency and violet the highest.[124]

Malebranche offered only a general treatment of reflection and refraction, drawing on Newton's empirical data. He thus provided a discussion of internal reflection (appearing for example in the transition from glass to air) treated in detail by Newton but no discussion of the reflection at the transition to a denser medium, a matter which Newton had given little attention. Malebranche's explanation of internal reflection bears a surprising resemblance to the hypothetical elucidation that Newton was to publish five years later. Malebranche assumes that in the transition from glass to air the ether vortices in the air exert a pressure that results in a reflection. He maintains that, given this pressure and using the same reasoning as for the collision of hard particles of ether, it can be deduced that the angles of incidence and reflection are equal.[125] Refraction is dealt with in a similar way. In his view the origin of refraction is that the glass contains fewer microvortices than the air, resulting in a differential pressure that causes the ray of light to change direction. He posits that frequencies remain constant in refraction, and renders this plausible by referring back to Newton's experiments, which conclusively show that colours are unchangeable.[126] This is at best a physical interpretation of Newton's theory of colours utilizing the frequency concept; at worst it is a circular argument.

Malebranche's sketchy treatment of optical phenomena accords with his general philosophical ambitions. His modest aim was to render acceptable the notion that his physical ideas on subtle matter – and on vortices – fitted into Newton's discoveries. Malebranche regarded these discoveries as being beyond question, calling his own theoretical ideas on optics 'conjectures' and 'insufficiently proven general insights'.[127]

There is very little discussion of other optical hypotheses in Malebranche's work. The most significant disagreements he mentions concern Descartes' views on the speed of light and on his hard globules of ether, as we have seen. Malebranche neither uses nor refers to Huygens' principle, whereas he does not engage the emission tradition, thus not explicitly rejecting it. When he writes about Newton, he does not reject the emission hypothesis; rather, the laudatory tone over Newton's optical experiments comes to the fore.

Malebranche constituted an important link in the medium tradition. He may have had no systematic, complete theory to offer, but he elaborated a number of the basic ideas put forward by Descartes and others, with sufficient thoroughness to provide a springboard for those coming after him. These successors used Malebranche's propositions as a basis for their own goal, which was the further conceptual and mathematical development of his conception. Moreover, they were faced with problems that Malebranche had not tackled, such as the rectilinear propagation of light and the dispersion of colours.

One of these successors was Johann II Bernoulli, who wrote only one treatise on optics, in 1736. He called it a physical and mathematical investigation into the propagation of light.[128] He was too modest, since he set forth the most complete medium theory of light and colours to be produced in the period between Malebranche and Euler. Bernoulli's treatise has not received its proper attention in the historical literature. The commentaries available on the physical aspects of Bernoulli's work are superficial and sometimes ahistorical.[129] In contrast, the mathematical side of his theory has been the subject of deeper study, yet sadly the connection with the physical context is to a large extent lacking.[130] A thorough, integrated study of Bernoulli's treatise is badly needed. Only an outline of his theory will be given in the present work, emphasising the resemblances and differences with regard to other representatives of the medium tradition.

Bernoulli adopts Malebranche's concept of microvortices, but unlike Malebranche, and in line with one of Newton's suggestions, he links colour and frequency by assuming the existence of periodically vibrating 'colour particles' distributed throughout the ether. Like Malebranche, Bernoulli adopts Newton's theory of colours, conceived as an empirical theory. Nevertheless, he goes further than Malebranche in elaborating on the consequences of these ideas of Newton's for a medium theory, examining colour dispersion and fits, for example. Finally, Bernoulli's mathematical elaboration of the subjects is wider ranging, especially with regard to the propagation of light.

The key to Bernoulli's theory is his ether model. He reproaches Huygens and Newton – in the latter case he is thinking of the ether as described in 1717 in the *Opticks* – for failing to furnish an explanation for the elasticity of ether particles. Following Malebranche, Bernoulli hopes to rectify this by assuming that the ether consists of an infinite number of tiny vortices, each seeking to extend itself as a consequence of the centrifugal force exerted by its parts; overall, this action creates the elasticity of the ether.[131]

So far there is nothing new here. However, Bernoulli introduces extra particles scattered throughout the ether. These are hard, extremely small, they vary somewhat in size, the distance between each of them is approximately a thousand times greater than their diameter, and a great many of these particles occur in every straight line.[132] They play a significant role in the propagation of light and the creation of colours. Bernoulli outlines the propagation of light in the following manner. The tiny particles are at rest as long as the pressure of the ether vortices on all sides remains the same. The balance is disturbed as soon as 'une force nouvelle' pushes one of the particles in a particular direction. The vortices between the particle being pushed and its closest neighbour are compressed. The compressed vortices cause the second particle to leave its place of rest in the line of propagation, followed by the third, and so on. In this manner a great but finite number of particles are displaced, all this taking place before the microvortices between the first and second particles have been compressed to their smallest size. Where this happens the microvortices once more extend, pushing all the particles simultaneously in the opposite direction. After a second thrust the whole process is repeated. The result is that the particles produce extremely rapid and extremely tiny longitudinal vibrations around an equilibrium state. Bernoulli gives the name of half 'light fibre' (*fibre lumineuse*) to the straight line stretching from the source of the movement to the end of the first compression, including a great number of particles.[133] Under the influence of an equilibrium disturbance with a specific frequency the particles, originally distributed in a random way, regroup in fibres containing particles of equal size. According to Bernoulli, the magnitude of the first particle determines the way in which the rest of the fibre is filled. Particles that are smaller or larger than the first one are unable to follow the original motion so easily and are expelled from the fibre. Bernoulli compares this process to the resonance vibrations of a string on a musical instrument, influenced by the sounding of another string lying close to the first and tuned in the same way.[134]

In his concept of a light fibre Bernoulli offers a medium theoretical conceptualisation of Newton's homogeneous ray of light. It may be that Newton provided the immediate inspiration for this model, since Bernoulli overtly compares his own hard particles with Newton's colour-making particles, although not without drawing attention to the differences.[135] Although Bernoulli does not say so, it is also conceivable that his hypothesis draws on that of de Mairan, who, when discussing the theory of sound, had posited – also by analogy with Newton's colour-making particles – the existence of *air* particles, able to transmit only one particular tone.[136]

Before giving a mathematical treatment of the way light is propagated, Bernoulli discusses the physical explanation for rectilinear propagation. He rejects Huygens' principle, although he considers it to be very ingenious. Bernoulli is not satisfied with Huygens' assurance that light affects our eyes most at the line crossing all the pulses perpendicularly (the pulse front), because it does not follow that no effect at all can be perceived in a direction slightly diverging from the straight line. Consequently Huygens' principle cannot explain the existence of sharply defined shadows, and Bernoulli appreciates a difference in this respect between sound and light, which otherwise strongly resemble each other. Sound travels not only in straight lines but also – though more faintly – obliquely, whereas the propagation of light (in homogeneous media) is always rectilinear.[137] It is remarkable that here Bernoulli does not broach the subject of the inflection or diffraction of light any more than Huygens had done, although they could (at least when viewed superficially) remove the objection to Huygens' principle. It is unclear whether Bernoulli's objection derives, directly or otherwise, from Newton. At all events, his criticism shows that Huygens' principle, and therefore his theory, is rejected by a natural philosopher from within the medium tradition itself for the same reasons as those customary in the emission tradition.

Bernoulli's explanation of the rectilinear propagation of light depends directly on the incompressibility of the colour particles. These can therefore, in Bernoulli's view, exert force only in the direction in which they are displaced, that is, in the line of propagation, not at the sides. This contrasts with a particle of air, which decreases in size in the direction towards which it is being compressed but increases its size in the plane lying perpendicular to it. This results in new, obliquely oriented 'sound fibres' that branch off from the 'main fibre', and also produce sound. Bernoulli does not clarify why the same effect does not occur with the ether vortices *between* the colour particles. He appears to be assuming that the ether vortices transmit a force in only one direction. In this way he is only shifting the problem to the micro-level of the vortices.[138]

Bernoulli introduces a new element into the discussion on rectilinear propagation. Newton, and implicitly Huygens as well, used essentially the same model for the medium of light and sound, thereby to a considerable extent reducing the scope for exploiting possible differences between the two phenomena. Bernoulli increases this scope by making a qualitative distinction between both of the media, but as far as I know, nobody elaborated his baroque version of this idea.

From that point in his treatise Bernoulli confines his attention to the longitudinal motion in a fibre. The laws for this movement are valid for both light and sound. In Bernoulli's view there is a far-reaching similarity between the motion of the string of a musical instrument and that of a fibre.[139] He compares the transverse displacements of such a string, fixed at both ends, with the longitudinal displacements of the colour particles in a light fibre. It is therefore possible that he is using the mathematics for standing waves, which seems strange to someone living in the twentieth century. Nevertheless, the use of this mathematics is consistent with Bernoulli's physical description of the propagation of an equilibrium disturbance in the light medium; at a finite distance the disturbance is 'extinguished' and all colour particles produce a vibration with the same phase and frequency.[140]

Bernoulli deals only briefly with the reflection of light. He believes the equality in the angles of incidence and reflection to be based upon the laws governing collisions. The ether vortices collide with the 'surface' of a body, but this neutral term leaves problems concerning the structure of bodies dangling in midair. Bernoulli also provided no discussion of internal reflection.[141] For refraction Bernoulli harked back to an article written in 1701 by his father, Johann I Bernoulli, in which the deduction of the refraction law had been turned into a problem in statics. His father's treatment was deliberately independent of any conception of the nature of light, being based on the equilibrium between two forces acting upon a mobile point in the same plane but along different lines.[142] It only remained for Johann II to show how the two forces in his theory of light come into existence. He identified these with the differing elastic forces of the ether, for instance, those in air and water. In water the ether vortices occupy a smaller volume, giving the ether therein a greater elastic power.[143] Johann I believed that equilibrium is reached if the ratio of the forces is in inverse proportion to the ratio of sin i and sin r (i = angle of incidence, r = angle of refraction). Johann II is able to interpret this in the following way for the transition from air to water:

$$\sin i/\sin r = \text{elastic force in water}/\text{elastic force in air}$$

Since he regards it as 'fort probable' that in refraction the frequency of the light ray remains the same, the velocity in water is greater than in air because of the ether's increased elasticity. He himself calls this a paradoxical and remarkable conclusion, because it runs counter to the widely shared idea that denser bodies have a retarding effect on the propagation of light. He mentions only Newton as an exception to this rule.[144]

Bernoulli deals somewhat tersely with colour phenomena since they are not part of his subject. Here we can see Bernoulli's position in relation to that of Newton and the emission tradition. He considers the heterogeneity of white light to be proven, and in his view the result of Newton's 'discovery' is that the modification theory of colours has come to be generally regarded as incorrect. Praise for Newton does not extend to his hypotheses on the nature of colours.[145] Bernoulli sees a heterogeneous beam as a bundle of homogeneous fibres that, in refraction, disintegrate into their constituent parts. He believes that dispersion results from a difference in velocity between the various fibres in a beam. Bernoulli does not tackle the problem of which colour has the greatest frequency.[146] Other colour phenomena are not dealt with, but Bernoulli makes an exception for Newton's rings. He considers Newton's explanation through the fits of easy transmission and reflection to be perfect of its kind, 'parce qu'il en déduit heureusement grand nombre de circonstances, qui toutes se vérifient par l'expérience'. One is tempted to expect that Bernoulli's enthusiasm for the fits is inspired by Newton's use of a hypothetical ether in order to explain the occurrence of fits. However, Bernoulli makes absolutely no mention of this suggestion of Newton's. On the other hand, he does perceive difficulties in accounting for fits within an emission schema. He cannot imagine how the fits, which occur in the very emission of a light particle, exist in a uniform medium. Bernoulli may have omitted Newton's ether hypothesis, which diminishes the seriousness of this objection, in order to enhance his own solution, for he thinks that the colour-dependent periodic fits can be identified with the colour-dependent periodic equilibrium vibrations of the particles in a fibre.[147] Although Bernoulli confines his comments to this general observation, this appears to be the first time that anyone working within the medium tradition has connected the frequency of light vibrations with Newton's rings. Later on Euler and Thomas Young present two different attempts to explicate this link both quantitatively and qualitatively. Only Bernoulli establishes a direct link with Newton's fits; Euler and Young will not adopt the concept of fits but will propose resonance and interference, respectively, as basic ideas.

Bernoulli's position within the medium tradition is a complicated one. He builds his own ideas on those of Malebranche while rejecting Huygens' theory. Despite the fact that he does not deal with such phenomena as diffraction, double refraction, and the colours of bodies, his is the most comprehensive theory for the period between Malebranche and Euler; nonetheless Euler repudiates several of Bernoulli's main ideas. Moreover, Bernoulli's theory makes it plain to us that the way in which periodicity and colour were eventually linked was not clear from the start. After Malebranche had introduced the general idea of relating periodicity and colour, all kinds of options were still open on both the ontological and the mathematical levels. Bernoulli's colour particles in an ether and the mathematics of standing waves provided only one of those options. The elaboration of the concept of periodicity in the medium tradition comprises a complex and fascinating historical process spanning a considerable period of time, which seems to have been unnoticed by historians of science so far.

In Bernoulli's work there is no strong awareness of an antithesis between emission and medium traditions. Newton's theory of colours is accepted on the empirical level as an adequate description of the phenomena. Newton's hypothetical ontology of colour particles is taken over as well, although in an adapted form. Bernoulli's criticism of the emission tradition is of a technical and detailed kind, the discussion of fits constituting an example of this.[148] Although Bernoulli rejected the emission hypothesis, he did not condemn it on the grounds of fundamental objections, as Euler was to do at a later date. This is all the more remarkable in view of the way in which Bernoulli discusses another question, addressing a biting remark to 'Mrs. les Newtoniens' because they posit a universal attractive force for which they have absolutely no intelligible (for this read 'mechanistic') cause. He makes an exception for Newton himself, indicating the place in the *Opticks* in which Newton suggests an ether as a possible explanation for gravitation.[149] The difference between Bernoulli's treatment of the 'Newtonians' and of the emission tradition shows that the link between 'Newtonianism' and the emission tradition was not overrigorous even in its opponents' eyes.

3. The relationship of the optical traditions

It has often been claimed that emission and medium traditions had a monolithic appearance, had a completely distinct evolution, and differed completely from each other.[150] The evidence presented above shows these claims to be false. The variations within each tradition have been pointed out repeatedly. In this section I shall argue that there was remarkable cross-fertilization, whereas rivalry was not always greatly stressed and was at times even entirely submerged in a synthesis of viewpoints. At the same time I shall be examining the extent to which the alleged conflict between 'Cartesianism' and 'Newtonianism' actually compelled the choice of either the medium or the emission tradition.

a. Exchange of concepts

As we have seen, in 1672 Newton regarded the 'emission or medium' question as less relevant than the question of 'matter or quality'. De Mairan claimed that light was either a material effluvium or a motion in an ethereal matter, unless one were prepared to take refuge in the idea of 'occult qualities'. For both natural philosophers the antithesis between Aristotelianism and mechanicism took priority. For as long as Aristotelian natural philosophy was regarded as a significant competitor, it was necessary for a mechanicist to confront this antithesis. Newton, with a characteristic 'perhaps', opted for the idea of matter and motion. De Mairan made the same choice in a much more definite way by writing of 'les systemes modernes'.

In spite of the growing conflict within the mechanistic camp between the emission and medium traditions, there was a good deal to unite the representatives of the two traditions, in addition to their self-conscious rejection of Aristotelian natural philosophy. The fundamental agreement between the mechanistic traditions in optics permitted an exchange of ideas on the ontological level, even though it was usually necessary to adapt these ideas to their new context. One example of this was Johann II Bernoulli's adoption of Newton's idea of colour particles. The latter's particles, emitted by a light source and being of variable size and therefore differing in their velocities of propagation, were transformed by Bernoulli into vibrating particles of variable size and therefore differing frequencies. De Mairan converted the colour particles into resonating air particles, thereby making the leap from optical theory to theory of sound.

It was also possible to go from one tradition to the other in a second way, using physical illustrations. Descartes' comparison of the refraction of light to the motion of a ball was adopted by Newton and

Table 2. *Ideas on the ratio of light velocities in glass and air, held in the period 1600–1750*

	Emission tradition		Medium tradition	
	17th century	1700–50	17th century	1700–50
$v_{glass} < v_{air}$	Maignan	de Mairan (1717)	Hobbes Huygens	Euler
$v_{glass} > v_{air}$	Newton	's Gravesande van Musschenbroek	Descartes	Johann II Bernoulli

supplemented with a mathematical attractive force. At first Newton applied the comparison in the deduction of the law of refraction in the same way that Descartes had used it, that is, as a physical illustration, where only the mathematical relations were of interest. Only later was Newton to transform the illustration into a physical model of reality, in which the ball became a light particle and the imaginary attractive force became an actual force.

Finally, mathematical techniques and formulae could be taken over apart from any ontological model or physical illustration. An example of this can be seen in the development of the 'kinematic tradition' in seventeenth-century optics. Emanuel Maignan adopted the essentials of Hobbes' mathematical deduction of the law of refraction, while the physical conception underlying this changed from a medium hypothesis into one involving emission of light.[151] This transformation was achieved in a smooth way, for one reason, because in both cases identical relationships were posited between velocities when the medium changed. Maignan, like Hobbes, assumed that the velocity of light was greater in glass than in air, thereby fulfilling a condition for adopting the deduction from the law of refraction. As can be seen from the diagrammatic summary of assumptions about the velocity ratio (see Table 2), there was no clear dividing line between the two optical traditions in the seventeenth century or the first half of the eighteenth, although it is perhaps telling that in the period 1700–50 there appears to be no eminent proponent of the emission tradition who accepted the idea that light gains velocity in the transition from glass to air. Nevertheless, until the middle of the eighteenth century there was no likelihood of the idea taking form that the determination of the velocity of light in water and air could be decisive in the choice between the emission and medium traditions. It seems as though we have come across a reason that the idea of such a crucial experiment was put forward only in the second half of the eighteenth century.[152]

Besides the occasional exchange of mathematical techniques and deductions, there was a defense of the general and fundamental identity between emission and medium traditions on the mathematical level. De Mairan, who no longer categorically rejected the medium tradition after 1717 but still preferred a theory of emission, defended what I call a bridging thesis thus: 'les loix de la tendance des corps sont les mêmes que celles de leur mouvement & de leur transport actuel'.[153] These words remind us of a similar statement by Descartes, but for him this functioned in a less fundamental way, that is, exclusively to justify his physical illustration of the ball in motion, whereas de Mairan made the proposition serve as a general 'bridge' between the emission and medium traditions. As a rule, de Mairan utilized the emission image in his deductions because in his view this contributed to the clarity and lucidity of the explanation.[154] However, the bridging thesis guaranteed that a proposition and its proof could be transferred from one tradition to the other. After 1717 de Mairan no longer wanted to choose a specific physical theory of light. His observations were of a mathematical kind, although on occasion he diverged from this path to consider physical problems. Provided that a physical theory of light was mechanistic and did not obviously conflict with experience, he did not want to exclude that theory as a potentially correct explanation of the phenomena.[155]

b. The contradiction between the two traditions

In 1717 de Mairan was still highlighting the contradiction between the emission and medium traditions. Within the confines of 'modern' physics, that is, the mechanical philosophy, he perceived two, and only two, fundamentally different answers to the problem of the nature of light. Both answers conformed to the mechanistic requirement that an explanation should only use matter in a particular form, order, or state of motion. However, the manner in which de Mairan posed the problem of the nature of light was less customary in his time than one would expect if there is any truth in the idea of a fight to the death between the emission and medium traditions in the last quarter of the seventeenth century. In Newton's comprehensive 1706 discussion of competing conceptions of light, he did not convey the impression that there were only two mechanist options available. In his first 'critical query' he distinguished between theories that involved modifications of the rays of light and those that supposed original and unchangeable properties of the rays. Only after that did he go on to reject the medium theories and to adopt an emission hypothesis. Among the authors treated in this study Euler was, in 1746, the first after de Mairan to formulate the choice equally incisively.

It is often believed that the choice between medium and emission traditions ran a parallel course in the first half of the eighteenth century, with a preference for either 'Cartesianism' or 'Newtonianism', if the choice was indeed not actually determined by this latter preference. This was an obvious idea to entertain, since Descartes and Huygens could be located within the medium tradition, and Newton within the emission tradition. Moreover, a firm internal coherence was ascribed to both currents of thought. Added to the contradictions on such 'fundamental' points as the status of forces and whether or not space was filled, this meant that natural scientists in the first half of the eighteenth century could, so it was believed, be characterised by dividing them into two different and rival camps, the 'Cartesians' and the 'Newtonians'.

This representation of the period has recently been criticised in the historical literature as being too crude and even misleading. It has been suggested that there were different orientations within 'Newtonianism', and that 'Cartesianism' experienced a development that has induced some historians to talk of 'neo-Cartesianism' at the end of the seventeenth century and beginning of the eighteenth. The existence and significance of other currents of thought – 'Leibnizianism', for example – has also been defended. What was perhaps even more important was the criticism of the suggestion emerging from the polar image that every scientific question would, broadly speaking, be decided on the basis of a choice between a handful of fundamental assumptions. This suggestion turned out to be incorrect for the theories of electricity and magnetism. There were researchers in these fields who clearly belonged in one or the other camp, where the explanation of gravitation was concerned, for example, but who often used the same kind of theories.[156]

Where the nature of light is concerned, an image emerges lying between the old polar viewpoint and the as yet unintegrated, unstructured image that replaced the former position. Viewed *logically*, there was no firm link between the 'Newtonians'-versus-'Cartesians' debate, on the one hand, and the choice between emission and medium traditions in optics, on the other. In practice, connections of this kind were sometimes made, to be used – whether consciously or not – as tactical arguments. Some examples will serve to elucidate this point.

'Newtonians' are expected by those who adopt the old historical image to accept the idea of empty space and thus opt for an emission theory, whereas 'Cartesians' are expected to support a medium theory because they suppose space to be filled. De Mairan's position in 1717, in which the idea of filled space is partnered with an emission hypothesis, is an example that runs counter to these close linkages. His explicit assurance that he repudiates the idea of the vacuum, along with all kinds of other 'Newtonian' conceptions perceived as being connected with the emission of light particles, at the same time makes it clear that a standpoint of this sort is unusual in his environment. He is thus compelled in advance to defend his ideas against the criticism that seeks to establish a link with the fundamental debate between 'Newtonians' and 'Cartesians'. Newton himself implicitly contests the polar linkages, when, in 1717, he defends the idea that an extremely rarified ether can fill space without constituting a hindrance to the planets in their motions. In this he is clearing the path for his own ether and at the same time clearing it for the ethers propounded by the 'Cartesians', provided that the latter also claim their ether to be extremely subtle and rare.

Where the problem of whether space is filled is concerned, the 'Cartesian' Johann II Bernoulli also drew no firm dividing line between two camps. He did lash out at 'the Newtonian gentlemen' who embraced the idea of a universal attractive force. This appears to constitute an argument that the old polar image was sufficient for the status of forces. From within this image one would expect 'Cartesians' to repudiate forces operating at a distance as occult qualities, and conversely that 'Newtonians' would refuse to accept ethers as an explanation for physical action. However, the polar view concerning the first half of the eighteenth century ignores the 'neutral' conception of the status of forces, in which forces are regarded as entities that actually exist but that still await an explanation. Their physical basis thus remains undecided. This conception is called 'Newtonian' as well – with greater justice – and was propounded by the 'Newtonian' 's Gravesande, for instance. Nonetheless, we encounter this viewpoint within the medium tradition, as well as in the emission tradition.[157]

In conclusion, it can be stated that exchanges of various kinds were possible between the emission and medium traditions in physical optics. Or more precisely, the coherence of the two traditions and the contrast between them were still emerging and growing in the first half of the eighteenth century. In particular, the alleged polar contrast between 'Cartesians' and 'Newtonians' only permeated optics to a certain extent: The contrast did not determine the choice between one tradition and the other in the discussion on the nature of light. Moreover, only a few authors propounded in an explicit way the antithesis 'emission versus medium' in its extreme form, that is, that there are two, and only two, possible conceptions of the propagation of light. Only with Euler was the contrast between the two optical traditions to acquire the form fitting for the old historical image of optics.

4

Euler's 'Nova theoria' (1746)

When Leonhard Euler published his treatise 'Nova theoria lucis et colorum' (A new theory of light and colours) in 1746, he made a contribution to the medium tradition in physical optics that was without parallel in the eighteenth century. The 'Nova theoria' constitutes the most lucid, comprehensive, and systematic medium theory of that century. The significance of Euler's theory can be gauged partly from the fact that it was so widely discussed. No earlier attempts to provide an alternative to the theories developing within the emission tradition had stimulated pens to the same extent as Euler's 'Nova theoria'. It is remarkable that a relatively short treatise published as part of a collection of articles on a range of subjects created such resonances, and all the more so when we compare this reaction with the limited response to Huygens' *Traité* – a complete book – and with the way in which Johann II Bernoulli's prize essay was virtually ignored. A partial explanation no doubt lies in the quality of Euler's work, while his authority also made it difficult to ignore his 'new theory'.

Euler was probably the most important, or at any rate the most fecund, exponent of mathematics and natural philosophy in the Enlightenment.[1] Although his accomplishments in mathematics and mechanics are generally known and acknowledged, Euler's contributions to optics have attracted little scholarly attention up to the present. However, his role in the discovery of achromatic lenses has been the subject of historical study.[2] This part of his work falls within the field of mathematical and technical dioptrics, in which the perfecting of lens systems takes precedence. His articles on this subject comprise four volumes of the collected works, two more volumes being needed for Euler's summary in book form, the *Dioptrica*. In contrast, the fifteen articles on physical optics, including the 'Nova theoria', were collected in one volume.[3] Euler himself placed just as high a value, or a higher, on his 'Nova theoria' as he did on his dioptrical works, in contrast to the neglect of this treatise in standard historiography. His letters show that the forty-five-page article meant a great deal to him. He drew the attention of many of his correspondents to it, asking some of them for their opinion of it, and he was always ready to answer questions on his theory of light. He frequently returned to his 1746 treatise, both in specialist articles and in the *Lettres à une princesse d'Allemagne*, intended for a wide public.[4]

Here we will be concerned with Euler's work in physical optics, especially the 'Nova theoria'. We shall be primarily concerned with three sets of questions in the discussion of Euler's treatise.[5] What place did Euler occupy within the medium tradition, and what was his relationship to his predecessors? What induced Euler to bring to a head the antithesis between emission and medium traditions, and what were his arguments for and against the two traditions? Finally, what is the actual content of his wave theory, and in what respect did he surpass his predecessors with that theory?

1. Euler's place in the medium tradition

It is tempting to regard Euler's wave theory as the connecting link between Huygens' pulse theory in the seventeenth century and the wave theories propounded by Young and Fresnel in the nineteenth. In this reading, it is Euler who continued Huygens' uncompleted work. If Huygens admitted in so many words that he did not know how to deal with colour phenomena, Euler brought these important phenomena into the realm of the medium tradition. Whether explicitly or implicitly, this is the current image of Euler's role in the development of the medium tradition.[6] This view, however, gives rise to an intriguing problem concerning the relationship between Euler and Huygens. Why is there no mention of Huygens anywhere in the 'Nova theoria', whereas there are references to Descartes, for example? Further, how can Euler talk of a *new* theory in the face of Huygens' *Traité*?

When one compares the two theories, the problem becomes more apparent than real, resulting from the Great Men view of history discussed in Chapter 1. If Young and Fresnel built upon Huygens' work in the early nineteenth century, then this historiographical view naturally leads to the assumption that Euler did the same thing in the eighteenth, since the correctness of Huygens' theory would certainly have been recognised by Euler. Yet if Huygens' and Euler's theories are compared point by point, the differences are more striking than the similarities.

The only significant similarity is that the point of departure for both theories is the analogy between sound and light, the likeness being, however, restricted to that part of the analogy that in effect posits that light is a pulse motion in an ethereal medium. There are many divergences between the two theories to offset this single, more or less obvious similarity that provides the justification for locating both Euler and Huygens within the medium tradition. Huygens' ether is entirely different in kind from Euler's subtle matter. The former accepted the idea of extremely elastic ether particles that were in contact but left small interstitial spaces; for the latter the whole of space was completely filled, the ether being an entirely homogeneous, fluid, subtle, and elastic matter. Huygens saw the propagation of light as the transmission of the extremely small displacement of the ether particles; Euler saw it as the propagation of a disturbance of equilibrium within the 'normally' uniform density distribution of the ethereal fluid. Huygens explicitly excluded the possibility of a periodic succession of pulses; in contrast, for Euler the concept of frequency is fundamental to his treatment of colours, and in this he extended the analogy between sound and light further than Huygens had done.[7]

The difference between the two theories is expressed most forcefully in Euler's attitude towards what is generally regarded as the central principle of Huygens' theory: that of the pulse front. The notion that a pulse front constitutes the envelope of a collection of secondary spherical pulses is completely absent from Euler's theory, whereas it forms the basis for Huygens' treatment of reflection, refraction, and double refraction. There is more than one possible answer to the question of why Euler did not utilise Huygens' principle. The most likely answer is also the most obvious, in light of the way the medium tradition developed in the first half of the eighteenth century. Huygens' principle had been dismissed as early as the beginning of that period; Euler was thus merely following in his predecessors' footsteps. As we have seen, the main drift of criticism, both within and outside the medium tradition, was that Huygens' theory was unable to provide an explanation for the propagation of light. Euler appears to agree with this criticism without actually mentioning Huygens himself, as we can see from a later passage in the 'Nova theoria', to be discussed below.[8] It is conceivable that Euler might have set aside his objections to Huygens' theory if he had devoted some attention to the refraction observed in Iceland crystal and to Huygens' explanation of this phenomenon. However, Euler does not touch upon the subject of double refraction anywhere in his work. A second reason for the absence of Huygens' principle in Euler's wave theory may have been that Huygens' principle, together with the hypothesis of periodic waves, conflicts with the rectilinear propagation of light. (Rectilinear propagation can only be explained with the addition of the idea of wave interference.) Following upon this, it has been suggested that Huygens rejected the periodicity of waves to avoid having to surrender his pulse front principle.[9] In this line of thinking, Euler could have done precisely the opposite; that is, he might have rejected Huygens' principle in defense of the idea of periodic impulses. Due to the lack of any evidence in its favour, this hypothesis has to be rejected. It is also looking too far from home. In Euler's time Huygens' theory was not sufficiently prominent to necessitate a thorough investigation. The theory had already been repudiated before it could be considered by Euler. Even if we assume that Euler examined Huygens' theory himself, the difference in assumptions about the ether are sufficient to make his theory unattractive to Euler.

Nevertheless, there is a grain of truth in the portrayal of Euler's wave theory as an extension of Huygens' pulse theory. It was certainly Euler who systematically incorporated the concept of frequency into the medium tradition, resulting in a theory that had an increased explanatory range compared with Huygens' theory. However, when interested in conceptual evolution, we are obliged, in this version of the story, to replace Huygens' theory with the subtradition of Malebranche and Johann II Bernoulli, adding the caveat that the introduction of the frequency concept should not be ascribed to Euler alone. It should be seen instead as a gradual process, acquiring its first recognisable form with Malebranche and reaching a provisional conclusion with Euler. If Euler is located at a point along this continuum, he is no longer the first Great Man after Huygens, but rather the one who completed a development that had acquired clear outlines with Malebranche, half a century before his time. Euler added his own ideas to those of his predecessors, systematically combining them in a new synthesis.[10]

Like his forerunners in the first half of the eighteenth century, Euler saw part of his task as the accommodation of Newton's theory of colours in the medium tradition. Yet he opted not for the hybrid solution proposed by Johann II Bernoulli, who posited the existence of separate colour particles in an otherwise homogeneous ether, but for a solution that was closer to Malebranche's pressure vibrations. However, Euler went further, also providing a mathematical treatment of the propagation of light pulses that draws heavily on Newton's theory of the propagation of sound, given in the *Principia*. Euler also had greater awareness of the problems attending the identification of colour and frequency. He tried to find answers to open questions concerning, for example, the continued existence of colours in the process of refraction, the production and composition of white light, refractive dispersion, and the colours of opaque bodies. Where the last subject is concerned he appears to have been inspired by Bernoulli, except that the latter's 'resonance particles' are located by Euler not in the ether, but in the coloured bodies themselves.[11]

Another new feature of Euler's 'Nova theoria' is the way in which he presented his ideas. The 'Nova theoria' begins with ample discussion of the arguments for and against the medium and emission traditions. This survey marks a transition point in the relationship between the two optical traditions. The drift of Euler's argument moves inexorably in one direction: No more hybrid theories à la Bernoulli, no more 'bridging' theories of de Mairan's kind. The time for barter or synthesis was past: The duel had begun.

2. Arguments for and against

Why did Euler preface the presentation of his theory of light with a discussion of the arguments for and against it? I believe that he was virtually forced to do so. If he wanted to succeed in convincing others of the superiority of his own theory, he had, at the very least, to remove the barriers erected in the preceding period that could have prevented acceptance of his own theory. These barriers were the objections to the medium tradition and particularly the argument on the rectilinear propagation of light. Euler had probably had the personal experience that nothing resulted from advancing a medium theory without a discussion of the pros and cons. For when he expounded his theory for the first time, in a lecture delivered to the Berlin Academy in 1744, he barely engaged the familiar arguments against the medium tradition.[12] The arguments in favour of a medium conception of light and against emission theories are dispersed throughout his exposition, and they are formulated categorically. Only the analogy between sound and light, an argument in favour of a medium theory, is given a detailed treatment and a conspicuous place in the opening section of the lecture.

Two years later, in the 'Nova theoria', the presentation was drastically altered, the first of the work's five chapters being given entirely to a systematic discussion of the pros and cons. Virtually all of the analogy argument disappeared, and the other arguments had been changed extensively. These differences, which will be discussed below, give rise to the suspicion that Euler changed the presentation of his theory in response to its lukewarm reception by his colleagues in Berlin. If his main aim in 1744 was to present his theory, by 1746 he also wanted to convince his public of the theory's validity.

Euler's tone in the 'Nova theoria' is self-confident and almost pedantic, the same conviction being virtually always present in his letters, articles, and *Lettres*. On only one occasion was he prepared to use his opponents' hypothesis of the emission of light matter as a fictional possibility permitting of a simple mathematical formula.[13] However, Euler did not express any genuine doubts about the correctness of his own theory of light. On the contrary, Euler's faith in his wave theory was so patent that one of his supporters felt obliged to point out to him that the theory ought to be presented 'mit weniger assurance'.[14]

The very first sentence of the 'Nova theoria' provides an example of Euler's self-assured style:

> Every sense perception occurs through contact, by which means a certain change in our body is produced; both reason and experience teach us this so clearly that there can remain no further doubt about it.[15]

Euler thus rejects the idea of action at a distance as causing the stimulation of the senses, without stating so explicitly, but the passage does not entail a choice in the way in which contact with our senses is effected. There is no problem where an object comes into direct contact with a sense organ, as in the case of touch and taste. However, when the objects are at a greater distance from us, as in the case of smelling, hearing, and seeing, Euler believes that there are only two possible ways in which bodies are able to stimulate the senses:

> . . . either . . . effluvia flow from these bodies, and encounter our sense organs, or they excite such motion in the surrounding bodies that this [motion] is propagated to our senses by all the intermediate bodies.[16]

Euler thus distinguishes between the propagation of emitted matter and that of a motion in a medium. For the first he provides the example of the distributions of smells by means of fine particles, and for the latter the example of the propagation of sound by movements in air.

Euler notes that opinions diverge on the explanation of sight. He cites Newton as having opted for the emission of matter, by analogy with the explanation for smelling. (Here Euler, like others before him, is ascribing a 'Newtonian' tenet in an unconditional way to Newton himself, whereas Newton had presented his ideas on the emission of matter in a hypothetical vein.) By contrast, Descartes and 'most other philosophers' favoured the idea of a motion in a medium, by analogy with the mechanism in the hearing of sounds.[17] (This is also less than the whole truth, where Descartes is concerned.)

With the aim of providing a decisive answer to the question of the nature of light, Euler announces that he will give a carefully balanced exposition of the arguments usually advanced in favour of both hypotheses. He accounts for this investigation of the arguments pro and con in the following way:

> Although . . . the best support for each theory lies in the perfect explanation . . . of all phenomena, in order to prevent the doubt that has been conceived at the very beginning from keeping the reader's mind in a state of uncertainty for any longer, I shall nevertheless take the trouble to demonstrate that the second conception [i.e., the medium conception] . . . is not only more probable than the other, but is also entirely in accordance with the truth.[18]

Two things strike us about Euler's way of introducing the problem of the nature of light. Although he restricts the number of possible theories of light by his condition of contact action, this has no direct consequences for the choice between emission and medium traditions. Even those supporters of an emission theory who make use of attractive and repulsive forces as the physical explanation of such phenomena as refraction and reflection are not affected by Euler's criterion, since these forces play no part in the contact between light particles and eye. Thus, at this point in Euler's argument emission and medium theories are equally acceptable. Therefore, Euler was one of those who saw the discussion on the nature of light as a fraternal quarrel involving 'only' the specific mechanism of the propagation of light.

The second feature to attract our attention in Euler's presentation of the problem is that he acknowledged two, and only two, possible solutions, thus creating an either/or situation in which, as we shall see, an argument against an emission theory is also an argument for a medium theory. As we saw in the previous chapter, Euler was not the first to narrow the choice down to two alternatives, but this presentation had previously been used in textbooks, with the exception of de Mairan's treatise. Euler was the first influential 'researcher' to pose the choice in such a black-and-white way and to place so much emphasis on it, devoting an entire chapter to it.

Euler discusses six arguments in all. The first of these supports his own medium theory, and numbers two and three constitute the reply to objections. In the obvious belief that attack is the best form of defense, he addresses problematic features of the emission tradition in the last three arguments.

1. The first is an analogy argument in favour of a medium conception. Euler observes that nature does not avail herself of effluvia in the distribution of sounds, but 'because of the rather large distances' utilises another method of propagation. It therefore seems very probable to him that nature also makes no use of effluvia for bridging the much larger distances occurring where light is concerned. He also believes that there are many similarities between hearing and seeing, whereas smelling and seeing are very different. Therefore, the propagation of light ought to display a much greater similarity to that of sound than to the propagation of odours. Although Euler ascribes considerable force to this argument, he recognises that

> . . . for those who are devoted to the rival conception [i.e., of emission] the consideration [of this argument] usually seems hardly worthwhile; it is for this reason that the very foundations upon which they attempt to base their opinion have preferably to be examined and enfeebled.[19]

I presume that this passage contains autobiographical elements; the grounds for this presumption becomes clear when we compare the analogy argument just reproduced with the opening section of the 'Pensées sur la lumiere et les couleurs', the lecture Euler had given two years before in 1744:

> Il y a un si grand rapport entre la lumiere et le son que plus qu'on recherche les proprietés de ces deux objets, plus on y decouvre de[20] ressemblance. La lumiere et le son parviennent à nous tous les deux par des lignes droites, s'il ne se trouve rien, qui empeche ce mouvement et en cas qu'il s'y rencontre des obstacles, la ressemblance ne laisse pas d'avoir lieu. Car comme nous voions[21] souvent la lumiere par reflexion, ou par refraction, ces deux choses se trouvent aussi dans la perception du son. Dans les echos nous entendons le son par reflexion, de la même maniere, que nous

> voions des images dans un miroir, la refraction de la lumiere est le passage des rayons par des corps transparants, qui produit toujours quelque changement dans la direction des rayons; la meme chose se trouve aussi dans les sons, qui passent souvent des murailles et d'autres corps, avant que de parvenir à nos oreilles: de sorte que les murailles et d'autres corps semblabes sont par rapport au son la même chose, que les corps transparans par rapport à la lumiere. Il n'y a aussi nul doute, que le son en passant par des corps propres à le transmettre ne change de direction, quoiqu'il soit bien plus difficile de s'appercevoir exactement d'un tel changement dans le son que dans le lumiere. Il est pourtant aisé de s'assurer[22] de cette deflexion du son, quand il passe par des murailles assez[23] obliquement, car alors on se trompe ordinairement en jugeant de l'endroit d'où le son est parti.[24] Une si grande ressemblance ne nous permet pas douter,[25] qu'il n'y ait une pareille harmonie entre les causes et les autres proprietés du son et de la lumiere, et ainsi[26] la theorie du son ne manquera point d'enclaircir considerablement celle de la lumiere.[27]

This is Euler's introduction to his lecture; he sets out his theory immediately after these observations. The construction of the introductory section is remarkably weak. As can be seen from the last sentence of the passage quoted, Euler's intention was to apply the theory of sound to the theory of light, and he defends this method in his introduction, in two stages. The first is the thesis of a 'harmony' between the causes and 'les autres proprietés' of sound and those of light. It follows that the theory of sound can clarify the theory of light. The second step is an argument for the validity of the first step, that is, for the existence of the analogy between light and sound. This argument is that sound, like light, is propagated rectilinearly and is susceptible to reflection and refraction. This claim is unfounded in the case of the propagation of sound. Concerning reflection, Euler mentions the echo without going into any detail. With regard to the refraction of sound, he first says that it is difficult to determine the extent of refraction with any precision, positing subsequently that where there is a sufficiently oblique incidence, the directional change of sound can be deduced from observation.

This was a weak 'argumentation', since all three properties Euler had imputed to sound were by no means obvious, except perhaps for reflection, and had certainly not been verified by experiments. The vague observation about refraction does nothing to fill this lacuna, and the rectilinear propagation of sound, the most controversial conception since Newton, is posited without any evidence.

To exaggerate, it can be claimed that Euler's argumentation in favour of the analogy between sound and light is in fact likely to have convinced his audience of the very opposite. We can find an indication that Euler's argument did have a cool reception, in the abridged account of the 'Pensées' given in the proceedings of the Berlin Academy. This summary was probably written by Jean Henry Samuel Formey. The last sentence of Euler's opening section in the 'Pensées', in which he concludes that the complete 'harmony' between sound and light cannot possibly be doubted, is replaced in the summary by a much more modest statement: 'Une pareille harmonie entre les effets semble en indiquer entre les causes'.[28]

Keeping in mind this picture of the reaction to Euler's 'Pensées', we can understand why he wrote, in the 'Nova theoria' of 1746, that his opponents attached little value to the analogy argument. This may also be the reason why Euler made substantial changes in the presentation of his arguments in 1746, formulating his analogy argument in the 'Nova theoria' very differently from the way he had presented it in the 'Pensées', and set out to convince his opponents that he was right, by undermining the basis for their ideas.

Euler seems to have performed this task with ill grace, something that can be inferred from the motivation underlying his investigation of the pro/con arguments, cited above from the 'Nova theoria'. In the light of the presumably negative reaction to the 'Pensées', this fragment also acquires an autobiographical significance. Euler says here that although his wave theory has to speak for itself, for those with minds given to preconceptions, he is willing to take the trouble of demonstrating that motion in a medium is more likely than emission of matter. Thus, Euler's classic text on the pro/con arguments was probably written with a provoked heart. Moreover, what actually happened, as we will see presently, was exactly what he did not want; that is, the full-fledged theory was judged mainly on the basis of preliminary arguments for and against general hypotheses and barely, if at all, on the basis of the specific qualities of his new wave theory of light.

2. Euler's second argument, he claims, is the most important one for the champions of the emission theory. It looks back to Newton and is based on the objections to the idea of a plenum. Sound cannot be

propagated where there is no air or other matter. Were light to be propelled in a way analogous to that of sound, the space between the earth and the sun ought to be filled with a subtle matter. Since planets and comets would encounter resistance from this matter, and because Newton had been unable to perceive any effect at all on the motions of the heavenly bodies, he had been, in Euler's view, compelled to posit that 'the heavens were free of any resistance and [were] consequently empty'.[29] According to Euler, here Newton was disputing Descartes' proposition that none of space is empty, and Newton's successors would regard this argument as unassailable. However, Euler thought it to be incorrect since it gave rise to anomalies.

The same who try to banish all matter from space, fill that same space with light particles emitted by the sun and stars. Since, Euler says (echoing the objection), they clearly believe that the resistance experienced by the heavenly bodies because of the presence of this matter is so slight that its effects are still not discernible after many centuries, the original objection to a medium hypothesis becomes invalid. After all, if the 'Newtonians' maintain that planets and comets can move unhindered through a space filled with emitted light particles, the argument against the existence of a light medium is rendered null and void, supposing that the resistance offered by this medium is no greater than that allowed for by observation. To this one should add (here Euler is getting into his stride) that the light particles have an extremely high velocity that 'exceeds all our imaginative powers'. Consequently those who believe that the space between the heavenly bodies is empty are compelled to acknowledge that this space is not only filled, but is also in a state of continuous motion.[30]

Here Euler cleverly emasculates an objection to the medium tradition. He produces a caricature of Newton's position, suggesting that Newton regarded it as proven that space was *totally* empty, whereas in fact, as we have seen, Newton makes an exception for rarified vapours, a very attenuated ether, and – this is mentioned explicitly – rays of light (for which read 'light particles').[31] By making Newton's theory too extreme, Euler creates a target, and he does not shrink from using Newton's own ammunition against him. As we shall see in due course, later commentators from within the emission tradition regarded Euler's reply mainly as a challenge to their own ideas on emission. The requirement of a virtually empty space – a requirement directed against a medium hypothesis – seemed to be no longer in question for those commentators. Euler's aggressive defense may have succeeded in attaining this, but it is not very likely given that the vacuum argument was already old-fashioned by 1746. Few supporters of the emission tradition were still using it as a weapon in the struggle against the medium tradition. Euler's reminder about Newton's criticism of Descartes is therefore likely to have been rhetorical, providing a prelude to the attack on the emission tradition with the aim of making that attack even fiercer. This scheme was very successful. Not only did later commentators interpret this argument as problematic for emission theories, but some even managed to find yet another objection besides the one connected with the filling of space with light particles. They regarded Euler's observation on the unimaginably great velocity of light particles as an additional argument against an emission conception, one that required an answer.[32] However, the cursory manner in which Euler recalled this well-known reasoning provided the majority of his opponents with an easy excuse for ignoring it.

3. The second objection to a medium hypothesis that Euler tried to defuse was one current in 1746 (in contrast to the one we have just discussed), and it was to remain a live issue until the nineteenth century. It concerns the rectilinear propagation of light, and like the first objection, it was derived from Newton. Euler refers to the place in the *Principia* in which Newton posits that pulses in an elastic medium – sound pulses, for example – entering a room through an opening, penetrate into all corners of that room. Since rays of light do not behave in this way, Newton concluded that light cannot consist of pulses propagated in a medium.[33] Euler's reply is surprising. The crux of his answer concerns the nature of the analogy between sound and light. In a complex argument backed up by a thought experiment, Euler tries to demonstrate that Newton was careless in his application of the analogy between sound and light and that his conclusion was consequently wrong.

Euler begins by admitting that, in a room with an opening through which sound enters, the sound certainly is more or less equally audible in all corners of the room. Nevertheless he contests Newton's explanation for this, that is, that the sound pulses spread past the opening into all the space in the room. Euler believes it to be an empirical fact that observation enables us to tell the location of a sound source, because we can judge this by the direction of the pulses reaching our ears. In a room with an opening we do not say that the sound is coming from the direction of that opening; consequently the sound is not spreading out into the room from the opening. It even follows that the sound must be audible everywhere in the room, even when the opening is closed. In Euler's view neither the bending of sound pulses nor an opening into the room is required for the perception of sound in all its corners. All that is necessary is that walls should allow sound to pass through them, a feature which Euler believes to have been confirmed by frequent observation. In this respect walls that allow sound to pass through them can be compared to translucent or transparent material.

Euler believes that Newton's attempt to demonstrate the dissimilarity between sound and light was therefore invalid. Newton was not making an honest comparison. Since in his optical research he had investigated the case of a room with walls that allowed no light to pass, he ought to have compared it with a space in which the walls were impervious to sound. Euler notes that an experiment with a soundproof room could establish the way in which the analogy between sound and light should be drawn. However, the technical problems in constructing a room of this kind are so great that he doubts whether the issue could ever be decided in this way. Euler nonetheless believes that he has demolished Newton's objection with this reply.[34]

Euler finds a second support for his viewpoint in a comparison of sound propagation in two different situations. If sound were to spread out laterally in a room, then it would have to do the same in the open. But, he continues, since the latter phenomenon has not been observed, sound must also be propagated in a rectilinear manner in a confined space. He provides no details of the indicated observations, but he does discuss a possible argument against them: The situation in the open differs from that in the room, because in the open air every pulse has a neighbour that prevents it from spreading sideways; if, on the other hand, a sound is propagated via an open window or the door of a room, there are no 'opposing' neighbouring pulses to the sides of the sound cone, resulting in lateral diffusion. Euler disagrees with this idea. In his view it is impossible to conceive how one pulse could prevent another from spreading sideways, for experience proves that various sound pulses do not disturb each other, but are propagated independently through the same part of the air.[35] Here it is once again clear that Euler regarded the propagation of sound as completely analogous to that of light: Both light and sound pulses were seen as propagated in rectilinear fashion in both the room and the open air.

In the next chapter of 'Nova theoria' Euler returns to the issue, after dealing with the mathematics of the propagation of pulses in a medium. He tries to demonstrate the absurdity of Newton's conception of pulse propagation. If it were true, Euler says, that pulses penetrate into the 'shadow', then they would for the same reason have to extend in a backwards direction as well; this conflicts with both theory and experience. The formulation of this observation reminds one of Newton's reply to Hooke in 1672, quoted earlier. Newton claimed that 'both *Experiment & Demonstration*' showed that pulses penetrate everywhere into the 'shadow'. As with Newton, so with Euler in his 'Nova theoria': The argument stopped at a categorical statement.[36]

Euler's singular views on the propagation of sound, and on the way in which the analogy with light ought to be drawn, stimulated a number of reactions in Germany. We shall analyse this discussion in the next chapter, where it will become clear that neither Euler's point of view nor any other hypothesis or experiment provided a solution to the problem of rectilinear propagation that would be acceptable to opponents of the medium tradition.[37]

4. Euler concludes his discussion of the two foregoing arguments with the observation that these had destroyed the foundations of 'Newton's' conception. Nevertheless, he had only answered part of Newton's objections to the medium tradition. Although defense occasionally turned into attack, the tenor of Euler's discussion was defensive; however, this was to change now.

According to Euler, his three arguments against the emission of light matter are very weighty and have never been answered, let alone refuted by strong counterarguments. Euler is either forgetful or badly informed on this point, because by 1746 two of the three objections had already been rebutted in a fairly convincing manner, as we shall see.

For Euler the first problem with the emission tradition was that, if light particles were streaming continuously from the sun's surface, the sun would gradually lose its matter. Although light can be pictured as extremely subtle, its density nevertheless remains finite, and the total amount of light rays emitted must ultimately comprise a considerable proportion of the sun's total matter. Euler clarifies this claim with a calculation, the first to be given in the 'Nova theoria', in which he computes the consequences of the sun's emanation of matter. He assumes that the sun has already been giving out light for five thousand years, in which period it has not become appreciably lighter in weight. (Euler regards the loss of matter as imperceptible if the total amount lost is no greater than one-hundredth of the sun's matter.) He thus calculates that the density of light matter in the vicinity of the earth is smaller than that of the sun's matter to at least a factor of 10^{18}. He concludes:

> Although I know perfectly well that the patrons of the emission of rays find nothing absurd in the magnitude of this stupendous number, yet I have no doubt that, precisely because of this, this conception will lose not a little of its probability for impartial judges.[38]

The factor 10^{18} is composed of two different elements. The first is the proportion between the density of the sun's matter and that of the solar matter immediately after it has been emitted by the sun. This proportion is approximately $2 \cdot 10^{13}$ or more. The second element contributing to the number derives from the spreading or flaring out of solar matter on its way to the earth. On the basis of this effect the relationship between the density of light near the earth is roughly $5 \cdot 10^4$. Taken together, the two contributions produce the impressive factor of 10^{18}.

Remarkably, this argument is concluded in a much less firm tone than the overture to the objections against emission conceptions would lead one to expect. If the three objections are called weighty in the introduction, this argument is suited only to winning over impartial judges. Further, if at first the objections are said never to have been answered, Euler now knows very well that defenders of an emission conception can accept the idea of an 'infinitely' small density for light, and the absurdity of huge numbers is a subjective matter. We have to conclude that Euler was in fact aware of the answers to his first objection but found them unconvincing. As stated, as early as 1717 de Mairan produced the two possible answers: either that light matter is extremely subtle or that matter returns to the sun. In general, the extreme subtlety of light was a much discussed theme, as we have seen in the case of van Musschenbroek, for example.[39]

5. Even if Euler implied that this argument was not a particularly compelling one, he felt that the second objection to an emission theory was decisive. He believed that if two or more rays coming from different directions met each other at one point, they would have to disrupt each other's motion. Yet experience shows that the disturbance that could in theory be expected does not in fact occur. For when a great number of rays pass through a small opening into a dark room or through the focus of a mirror or lens, one observes absolutely no change of direction where (according to an emission hypothesis) there ought to be repeated and violent collisions between the rays. To Euler this contradiction between experience and an emission hypothesis

seemed a very powerful argument for overturning a hypothesis of this kind.[40]

Euler ignores, or is not aware of, the proposed solution to this problem of crossing rays that had been given six years before the publication of the 'Nova theoria'. In 1740 Johann Andreas Segner had stated that a ray of light ought to be regarded not as a continuous stream, but rather as a series of loose particles with large intermediate spaces. Because the impression of light caused by a particle lasts for some time, we do not perceive that the flow of particles is interrupted. Segner gives an experimental estimate of the persistence time, the distance likely to exist between two light particles. In this he uses a phenomenon that Newton had also mentioned.[41] When one draws a circle with a glowing coal at a certain speed, one sees a complete ring of light. At the minimum velocity needed to obtain a fully lighted circle, the second light impression apparently begins at the very moment that the first ceases. Segner believes the persistence time to be half a second, but he takes a tenth of a second, for the sake of certainty. This is also the time interval between the arrival of two successive light particles. The distance between two of these particles is thus around thirty thousand kilometres, or five times the earth's radius, in Segner's words.[42] Given this spatial distribution of the particles in a ray of light, the crossing of light rays presents no problems. This disposes of the objection, yet Euler raised the matter again six years later.

The fact that Euler either did not know Segner's treatise or ignored it should not be held too much against him, since Segner's contribution, and his idea along with it, seems to have been forgotten by virtually everyone. Thus, in 1752 Thomas Melvill could advance the idea, as if it were his own, that light particles follow one another at a very great distance. It was not until 1768 that John Canton estimated the distance between the particles, employing, like Segner, the idea that the eye works with a persistence time. In recent historical literature the impression given is that Melvill and Canton were the first to have answered the objection concerning the crossing of light rays, in the manner described above. Here, too, we appear to have insufficient knowledge of earlier writers, since Segner's treatise is discussed in Rosenberger's history of physics, published in 1884. Rosenberger's work seems to have fallen into the same obscurity as Segner's. History repeats itself in the writing of history.[43]

6. The third objection to the emission tradition concerns the composition of matter. In Euler's view an emission hypothesis produces a

problem about the structure of transparent materials, his reasoning being as follows. First of all, the transparency of materials such as water and glass can only be accounted for by the idea of rectilinear pores or paths, along which the light particles can move freely and without impediment. He adds the observation that rays can pass in all directions through the material. He concludes from both these premises that transparent bodies ought to have straight paths in every direction. But however the structure of the material may be conceived, there can never be a free passage for the light particles in all directions, or – as Euler had said two years earlier in the 'Pensées' – there is no place for matter in a transparent material.[44]

The problem Euler touches upon here had already been diagnosed in the emission tradition, and by none other than Newton himself. However, it had not been solved, as the previous two problems had been, at least in principle. To a large extent Newton's treatment of the transparency of bodies resembles that of Euler (a factor justifying the supposition that Euler had Newton's discussion in mind when he was posing the problem). In Newton's view as well, a transparent body would have to have a large number of straight pores, but he regards his work as accomplished if he can make a reasonable case for the idea that a body contains a great many empty spaces. He believed that the problem could be posited as follows. The particles of a body are distributed in such a manner that they leave as much space free as they themselves occupy. The particles themselves consist half of smaller particles and half of empty pores. These smaller particles in turn consist of even tinier particles and of empty space, and so on, until we encounter solid particles without pores. If the composition of matter is viewed in this way, the quantity of empty space and, with it, the number of pores along which the light particles can move without impediment can, according to need, be increased by enlarging the number of particle classes, that is, by increasing the volumetric ratio between void and matter.[45]

With this 'solution' Newton poses the problem of transparency in a vaguer way than Euler, who emphasises that not only must there be a great many empty pores, but these pores must also satisfy the specific conditions in that they have to be rectilinear and found in every direction. Consequently Newton's explanation of the transparency of materials was not the final word from within the emission tradition. Bošković and Priestley, among others, expressed dissatisfaction with the conception of matter underlying Newton's explanation. Bošković supposed in his theory of matter, known as point atomism, that matter is composed of points forming the centres of a force. If force is plotted against distance from the point, the force curve begins with repulsion, asymptotically going to infinity for a distance approaching zero. The curve crosses the horizontal axis a number of times with increasing distance from the force centre, ending with a part representing the attractive gravitational force.

The first repulsive part of the force curve is particularly important for the optical properties of a material. An individual point atom is completely penetrable to light particles, because of their extremely high velocity, apart from a very small area round the centre. Interaction between a material body consisting of a huge number of point atoms and a light particle is to a large extent determined by the manner in which the point atoms are arranged. In Bošković's view the forces within a homogeneous medium cancel one another out, permitting a fast-moving light particle to travel through the medium without impediment, or to meander somewhat, but in effect to follow a straight path. In a nonhomogeneous medium, or on the borders of two different homogeneous media, the forces do not counterbalance one another, and the light particle is thus diverted from its straight path, the result being opacity, reflection, or refraction.[46]

Priestley partially adopted Bošković's theory in his own survey of optics published in 1772. Euler's formulation of the transparency problem was, in Priestley's view, a reason for supporting Bošković's theory rather than Newton's conception of matter.[47] In this case Euler's shafts were obviously well aimed and had some effect, unlike his first two objections to the emission tradition.

Euler concludes with the observation that he could mention further difficulties attending an emission conception. In his view, however, the most essential propositions supporting this concept are overturned by the considerations discussed above. He repeats that the best support for his theory does not reside in refutation of the opposing conception, but in a complete conformity with the phenomena. He states that the remainder of the treatise will not only make it clear to the reader that his theory satisfies this demand, but also remove any lingering doubt over the composition of bodies, and it will be shown to be a theory that accords best with the simplicity of nature.[48] In contrast to Euler's explicit aims, his eighteenth-century readers, both friend and foe, were to decide the issue on the nature of light using the kind of evidence brought forward by Euler in the first chapter of 'Nova theoria'. In doing so, they concentrated on arguments contra emission and medium theories, largely ignoring the direct evidence in favour of the theories, in particular the remaining four chapters of Euler's treatise. Euler's reference to the doubt concerning the composition of bodies obviously alludes to his last objection to the emission tradition. The notion that his medium theory reflects the simplicity of nature refers to the fact that Euler regarded the ether as not only the medium in which light was propagated, but also as the cause of cohesion, electricity, magnetism, and gravity. Thus, a single ether explains a wide range of phenomena.

3. The content of the theory

Not only had Euler, in the first chapter of his 'Nova theoria', contradicted the arguments against the medium tradition; he had also raised a number of difficulties regarding the emission tradition. It now seemed appropriate to present the details of his wave theory of light, and he did so in the remaining four chapters of the treatise. In Chapter 2, which deals with the 'formation and propagation of the pulses', Euler first provides a reasoned repudiation of Descartes' model of the light ether. He subsequently gives a mathematical description of the propagation of a pulse and an estimate of the elasticity and density of the ether (see Section a, below). The third chapter, 'On the succession of the pulses and on rays of light', provides the transition from pulse theory to wave theory. Once more Euler takes up his position against an earlier conception, de Mairan's hypothesis on 'resonance particles'. He then advances his basic explanation of colour phenomena, and a wave theoretical interpretation of Newton's theory of colours. Even though Euler lavishes great praise on Newton for his discoveries in the field of colour studies, he does not adopt his doctrine in unaltered form (Section b, below). The fourth chapter, entitled 'On the reflection and refraction of rays', covers more than promised. Reflection and refraction are indeed dealt with, but they are followed by an extensive discussion of refractive dispersion and the composition of white light (Sections c and d, below). The final chapter, the fifth, deals with 'luminous, reflecting, refracting and opaque bodies'.[49] Of special interest are Euler's observations on coloured bodies and on what I call undercolours and overcolours (Section e, below).

Euler's treatment of colour phenomena, which is based on the analogy between light and sound, comprises the most innovative part of his theory. Formey perhaps perceived this better than Euler himself did, saying in his summary of the 'Pensées': 'Ce qui'il y a de particulier à l'hypothese de Mr. *Euler*, c'est son parallèle entre le son & la lumière'.[50]

a. Propagation of a pulse

At the very beginning of Chapter 2 Euler makes it quite clear why he does not subscribe to Descartes' ideas on light. According to Euler, Descartes filled all of space with particles of the second element, globular and completely hard. Consequently Descartes thought that the propagation of light was an instantaneous phenomenon. If he had known, Euler says, that light moves forward within a finite time, he might have assumed that the ether particles do not touch each other, but are separated by a small distance. He would then have been able to explain the finite velocity of propagation. However, in Euler's view there is an insuperable objection to this modified hypothesis. If ether pulses were to be propagated rectilinearly in a given direction, it would be necessary for the centres of the globules to lie on a straight line in that direction. However, this assumption would be incompatible with principles of geometry; that is, it would be impossible to arrange the ether globules in such a way that their centres lay on straight lines in every – arbitrarily selected – direction. Thus, for Euler this second, revised ether model was also inadequate.[51]

We can regard Euler's rejection of the second ether model as an implicit repudiation of Huygens' ether, and his theory with it, for the second model bears a strong resemblance to Huygens' ether. As we have seen, this consists of fairly hard particles, possibly spherical in shape and touching each other. In the propagation of a pulse the particles are somewhat crushed together, a process that takes time and results in a finite velocity for light. The spaces between the particles in the modified Descartes model have the same function as the incomplete hardness of Huygens' ether particles. In addition, Huygens' model is equally vulnerable to Euler's geometrical objection. This means that Euler's criticism of the revised Descartes model also applies to Huygens' ether model. Although there is no direct evidence, a plausible assumption is that Euler had both models in mind when formulating his dissent. This would suggest that Euler rejected Huygens' theory (as others within the medium tradition had done earlier on) because of its problems in relation to rectilinear propagation.

By now we know that Euler's ether is different from that of Descartes and Huygens, but what did his ether look like? In the 'Nova theoria' Euler claims only that the ether is a subtle and extremely elastic fluid.[52] He is more forthcoming in other places, particularly in the *Anleitung zur Naturlehre*, published posthumously but probably dating from the years 1756–8.[53] There Euler distinguishes two fundamentally different kinds of matter: ordinary, 'coarse' matter and subtle matter, or ether. The elasticity of coarse matter is based on the change in the way in which the component parts are arranged in order: Its solid parts cannot be compressed; only the pores can grow or shrink. The density of the smallest particles of coarse matter is immutable and the same for all substances. Thus, the difference between gold and silver lies not in the kind of matter but in the arrangement of qualitatively identical particles of matter. In contrast, subtle matter contains no solid or smallest particles and is completely homogeneous, and its density is variable, depending upon the actual compression of subtle matter itself. The density of subtle matter is always much less than that of coarse matter. Euler states that there is a natural density for the ether, but that it is not found in this state. The ether is compressed to a high degree, a fact that explains its extreme elasticity. Euler leaves room for the possibility that there are different kinds of subtle matter with various densities. If there are indeed more kinds, he nonetheless prefers to call them by the general name of subtle matter.[54]

Returning to the 'Nova theoria', we see that Euler continues with the main subject of Chapter 2: the mathematical description of the propagation of light. He uses Newton's treatment of propagation within elastic media, particularly air, in the second book of the *Principia*.[55] He reformulates Newton's geometric treatment in algebraic form and in passing clarifies several obscure points in Newton's argument. Euler emphasises that the propagation of pulses in an elastic medium is a subject that still presents difficulties, necessitating the use of conjectures along with established principles.[56] Euler's ultimate aim is the deduction of the formula for the velocity of light. It should be remembered here that in Chapter 2 Euler confines his attention to the propagation of a single pulse. It is only in the subsequent chapter that he tackles the periodic nature of light.

Euler begins with an outline of the situation in the following way (see Figure 7). On the line AO the points A, B, and C are displaced in turn to a, b and c. The velocity of the pulse is v, and $AB = BC = L$.[57] Supposing that at moment t the pulse is in B, then the pulse was in C at moment $t - (L/v)$. The extremely small equilibrium movements of A,

A B C a b c O

Figure 7. Euler's view of the formation of a pulse (Euler, 'Nova theoria', 12).

B, and C are taken to be harmonic (with angular velocity m and maximum displacement α); that is, the equations for the displacements $Aa = d_A$, $Bb = d_B$ and $Cc = d_C$ are

(1a) $d_A = \alpha \{1 - \cos m (t + L/v)\}$

(1b) $d_B = \alpha (1 - \cos mt)$

(1c) $d_C = \alpha \{1 - \cos m (t - L/v)\}$

Euler now wants to calculate the force at work at point b resulting from the difference in the density of the 'particles' ab and bc.[58] Here he starts out from the expressions

(2a) $ab = L + d_B - d_A$

(2b) $bc = L + d_C - d_B$

Substituting (1) into (2), he obtains

(3a) $ab = L - \alpha \cos mt + \alpha \cos m (t + (L/v))$

(3b) $bc = L + \alpha \cos mt - \alpha \cos m (t - (L/v))$

Now he writes out the third term of these two expressions for ab and bc in factors:

(4a) $ab = L - \alpha \cos mt + \alpha \cos mt \cdot \cos (mL/v) - \alpha \sin mt \cdot \sin (mL/v)$

(4b) $bc = L + \alpha \cos mt - \alpha \cos mt \cdot \cos (mL/v) - \alpha \sin mt \cdot \sin (mL/v)$

At this point Euler uses approximations for $\cos (mL/v)$ and $\sin (mL/v)$ with the justification that α and therefore L are very small. Euler needs only the approximation equation for the cosine, as we can see when we calculate the quantity $bc - ab$, the only one Euler subsequently needs, with (4):

(5) $bc - ab = 2\alpha \cos mt \cdot \{1 - \cos (mL/v)\}$

Supposing that $\cos (mL/v) \simeq 1 - (m^2L^2/2v^2)$, it follows that

(6) $bc - ab \simeq (\alpha m^2L^2/v^2) \cos mt$

Euler reaches this equation by making direct use of the approximation equations for sine and cosine in (4).

Now let us follow Euler's argument again. In accordance with Hooke's law, to which there is no explicit reference, he posits that in a compressed portion of the ether, the 'elastic force' is proportionate to density, k being the proportionality constant. Therefore, he obtains for the force ΔE at point b resulting from the difference in density $\Delta\rho$ between ab and bc, with the aid of (6) (D is the normal density):

$$(7) \quad \begin{aligned} \Delta E &= k \cdot \Delta\rho = kD\,(L/ab - L/bc) \\ &= kDL \cdot \{(bc - ab)\,/\,(ab \cdot bc)\} \\ &\simeq \alpha kDL \cdot \{(m^2L^2)\,/\,(v^2 \cdot ab \cdot bc)\} \cdot \cos mt \\ &\simeq \alpha kDL\,(m^2/v^2) \cdot \cos mt \end{aligned}$$

In the last step we use $ab \cdot bc \simeq L^2$. After this, Euler constructs the equation of motion in a rather circumstantial way. In an abbreviated and adapted form this amounts to the following: By means of equation (1b) the velocity and acceleration at point b can be expressed in the following way (v_b is the time derivative of d_B, v_b' the time derivative of v_b):

$$(8a) \quad v_b = \alpha m \sin mt$$

$$(8b) \quad v_b' = \alpha m^2 \cos mt$$

The equation of motion of the 'particle' around b with a mass of ρL is

$$(9) \quad \Delta E = \rho L v_b'$$

Substituting (7) and (8) into (9) and using the approximation $\rho \simeq D$, we obtain

$$(10) \quad (\alpha kDLm^2/v^2) \cdot \cos mt \simeq DL\alpha m^2 \cdot \cos mt$$

Comparison of the left- and right-hand terms and the substitution of the approximately equal sign by the equal sign produces $k = v^2$, or

$$(11) \quad v = k^{1/2} = (E/D)^{1/2}$$

Here E is the 'elasticity' of the ether, and D its equilibrium density. This is the same formula as Newton's.[59]

From equation (11) Euler draws the well-known conclusion that only the density and elasticity of the medium determines the propagation velocity of a pulse. The exact motion of the particles round their equilibrium position and the strength of the force setting the medium into motion have no influence on the velocity. In Euler's view the independence from the pulse size is confirmed by the observation that loud and quiet sounds are propagated at the same velocity.[60]

The velocity of sound can also be calculated by means of equation (11). Euler had done this at an earlier date, and he now repeats the calculation, finding a value that is more than 10 per cent less than the measured value. Newton had attributed the difference to the volume of air particles themselves and the presence of 'vapours'. Euler had rejected this suggestion in 1727 in his *Dissertatio de sono*. His own suggestion, in the 'Nova theoria', did not solve the problem either: The discrepancy between the measured and calculated velocity of sound was to remain unexplained throughout the eighteenth century.[61] Euler seeks the answer in the approximations he used, indicating that in the derivation of equation (7) the product $ab \cdot bc$ is set equal to L^2, but that in reality $ab \cdot bc$ is somewhat larger than L^2, and that the elastic force and thus the velocity of the pulse must therefore be larger than that calculated by means of equation (11). He makes no comment on whether this correction is sufficient to remove the discrepancy between theory and experiment. He was probably uncertain about this, as is clear from his observation that the theoretical value resulting from (11) was sufficiently in agreement with experience to entertain no doubts about the proportionality of the velocity to $(E/D)^{1/2}$.[62]

Euler returns from the air to the ether and sets himself the task of estimating the density and elasticity of the light medium, taking equation (11) as his starting point. Towards this end he introduces the following definitions:

(12a) $1/r = D_{ether} / D_{air}$

(12b) $e = E_{ether} / E_{air}$

Here r and e measure the relative 'rarity' (inverse of density) and elasticity of the ether. According to (11) and (12), the following equation is valid for relative 'rarity' and elasticity[63]:

(13) $r \cdot e = (v_{light} / v_{sound})^2 = 387{,}467{,}100{,}000$

I will round this off to 3.875×10^{11} from here on.

Euler tries to estimate the minimum values of r and e. For the definition of r_{min} he refers to another article in the collection in which 'Nova theoria' originally appeared. This paper addresses the problem of the resistance experienced by the planets in the ether, and Euler repeats the results in his 'Nova theoria': there is no 'noticeable' change in planetary motions for $r \geq 3{,}875 \times 10^8$.[64]

In Euler's view the ether is not only the medium in which light pulses are propagated, but also the origin of the cohesion of bodies: An object does not fall to pieces of itself, because the pressure of the ether holds it together. The extra function Euler attributes to the ether enables him to determine the elasticity of the ether. He equates this with what is called absolute cohesion, which in the eighteenth century was measured experimentally. An elongated object, a thread for example, was stretched lengthwise by applying a force, usually obtained by hanging a number of weights on the thread. It was posited that, at the moment in which the thread breaks, the force being applied is equal to the absolute cohesion of the thread. Van Musschenbroek carried out the best-known experiments in Euler's time. In his *Dissertationes* of 1729, van Musschenbroek gave the measured values for elongated parallelepipeds with a square base (mostly of wood) and for metal wires. He incorporated the most important results into his textbooks, in the chapter on cohesion.[65] In the 'Nova theoria' Euler used experimental results on the absolute cohesion of materials, without indicating his source, which was probably van Musschenbroek's measurements.

Euler posits that at the moment of breaking the cohesive force exerted by the ether F_{ether} is as great on the base area of an object (a wire, for example) as the force F_{weights} deriving from the weights hanging on the object:

(14) $F_{\text{ether}} = F_{\text{weights}}$ (at the moment of breaking)

He also says that the forces of air and ether are proportional to their elasticities:

(15) $F_{\text{ether}} / F_{\text{air}} = E_{\text{ether}} / E_{\text{air}} = e$

Substitution of (14) into (15) produces:

(16) $e = F_{\text{weights}}(\text{in breaking}) / F_{\text{air}}$

Euler now calculates the relative ether elasticity e from the measured values for F_{weights}(in breaking), in the case of a small cylinder with a base area of 10^{-6} square feet. For this surface Euler finds $F_{\text{air}} = 2.24 \times 10^{-3}$ pounds (he uses the pound as his unit of mass and force). In this case equation (16) is converted to

(16a) $e = (1{,}000/2.24) \cdot F_{\text{weights}}(\text{in breaking})$

In experiments with various materials it was invariably found, Euler states, that the relative elasticity of the ether e was much less than 1000, that is, that F_{weights}(in breaking) was always smaller than 2.24 pounds. Van Musschenbroek does indeed give 0.8 pounds (gold) as the highest value and 0.05 pounds (lead) as the lowest for a cylindrical metal wire (values converted to Euler's diameter).[66] Van Musschenbroek's later tables contain higher values than 2.24 pounds for silver and iron wires. In 1746 Euler had already expressed some doubts about van Musschenbroek's earlier results, pointing out that the strongest materials had not been used in these experiments and that the ether did not press on the whole ground area, only on the parts that were impenetrable. In this way he qualifies his conclusions on the size of e, for both considerations imply that the elasticity of ether is greater than the values 'measured' in the experiments. Consequently, Euler concludes that $e_{min} = 1{,}000$ is a good value.[67]

Taking together the minimum values for relative elasticity and rarity, Euler arrives at the value resulting from the ratio of light and sound velocities, as set out in equation (13):

(17) $r_{min} \cdot e_{min} = 3.875 \times 10^{11}$

The identity of the two values in the equations (13) and (17) may not have been a total accident; nevertheless, it is a remarkable outcome. Euler's method for arriving at this result is even more extraordinary. He estimates r_{min} and e_{min} by a method that is independent of the theory of light, subsequently comparing their product with the value resulting from his optical theory. In principle, this procedure makes it possible to test the theory on the (light) ether. In 1746 there was no reason for Euler to think that his argument was mistaken, since the results obtained by the two different methods accorded surprisingly well with each other. Further historical research is necessary to determine whether new data on planetary motions or the cohesion of bodies induced Euler to adjust his ether theory.[68]

b. Colours: Pulse sequences

In the first chapter of the 'Nova theoria' Euler defends a medium conception of light; in Chapter 2 he outlines a pulse theory; in Chapter 3, which we shall be dealing with here, he provides the basic concepts for a wave theory. He makes the transition from a pulse to a wave theory very gradually, dealing first with the mutual influence of two successive pulses.

Let a pulse bring part of the medium from its state of equilibrium. In Euler's view two possibilities arise from the arrival of a second pulse in the same place as the first. The first possibility is that the part of the medium involved here is at rest again, and the effect of the second pulse is precisely the same as that of the first. The second possibility is that this part of the medium is still in motion, which 'disturbs' the effect of the second pulse. Euler states, without furnishing any direct arguments in support of his idea, that the first possibility is the most likely.[69] He considers that our experiences with sounds confirm this independence of the pulses. Low and high tones with low and high frequencies, respectively, have the same velocity. He perceives only one circumstance under which the pulses are no longer independent, that is, where the frequency is extremely high and the pulses follow upon each other so rapidly that they have a mutual influence. He elaborates on this idea in his later treatment of dispersion.[70]

Euler then discusses de Mairan's rival theory on the nature of elastic media. As we have seen, in his theory of sound de Mairan postulated air particles having differential elastic forces and transmitting tones of only one particular frequency. Similarly de Mairan assumed the existence of 'resonance particles' in the light medium, the ether. Euler's comments on de Mairan begin with the same rather arrogant turn of phrase we have already encountered. Euler claims that his own theory has already been thoroughly established and the remainder of his treatise will consolidate this certainty – this constituting sufficient proof of the incorrectness of de Mairan's concepts. However, he is anxious to leave no doubt in his readers' minds and therefore provides an explicit rebuttal of de Mairan's ideas. Euler's main objection is that de Mairan's medium cannot 'exist'. If resonance particles have different elastic forces, the most elastic particles will expand, thereby compressing other, less elastic ones, until all the particles reach the same degree of elasticity. (Here Euler does not deal with Johann II Bernoulli's 'fibre conception' of the light ether, which bears a likeness to de Mairan's 'resonance conception' and with which Euler was familiar.[71])

After this introduction, Euler engages his real subject, the propagation of light pulses. He sketches the propagation of light as shown in Figure 8. A is a body in vibratory motion, thus creating pulses in the ether. The distance between two pulses is indicated by d.[72] If $AB = BC$ we have an 'isochronic' vibration; that is, the pulse distance d is constant. For 'nonisochronic' vibrations $AB \neq BC$, or the pulse distance d is not constant.[73]

It is significant that Euler does not introduce the term or concept *wave*length here; he talks instead of the distance or time interval between the pulses. Euler regards individual pulses as primary, and the sequence of pulses and their combination as secondary. Our modern concept of wavelength reverses this relationship: the whole thing – the wave – takes precedence, and the separate units – pulses, wave crests, and so on – fulfill a derivative role. As often happens in the history of science, it is the new – in this case Euler's *wave* theory – that is dependent on the old where terminology and conceptions are involved, that is, the pulse concept of sound and light. Strictly speaking, it would therefore be better, where Euler is concerned, to talk of a 'periodic pulse theory' rather than of a wave theory.

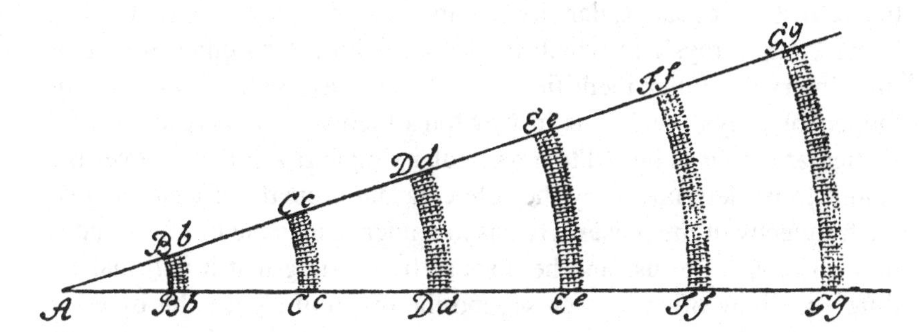

Figure 8. Euler's view of the propagation of a pulse sequence (Euler, 'Nova theoria', 22).

The concept of pulse is given no explicit definition. The most likely interpretation is that a pulse is the state of motion within that part of the ether lying within the circle sector *BbBb* or *CcCc*, and so on, the width of the circle sector (*BB*, *CC*, etc.) being variable (see Figure 8). A ray of light is defined as the imaginary line that cuts all pulses (in our words, pulse fronts) vertically, for example, the lines *AG*. In Euler's view the nature of the ray consists of two parts: (1) the direction of the ray, (2a) the 'frequency of the pulses' if the vibrations of the light source are isochronic, and (2b) the 'inequality of the time intervals between the pulses' in the case of nonisochronic vibrations.[74] Later in Chapter 3 Euler discusses the problem of whether the force or strength (*violentia*) of the pulses is part of the nature of the ray. He believes this not to be an essential property of the ray, since when the strength changes, only the 'liveliness' of the light is altered, not the colour. In the discussion of reflection in Chapter 4 Euler states that the direction of a ray is not part of its true nature. It is thus clear that frequency, or the irregularity of time intervals, is the only essential property of a light ray.[75]

After giving these definitions, Euler considers the way in which the pulse distance d can be determined for isochronic vibrations. In this case he gives the well-known relation among pulse distance d, frequency f, and pulse velocity v:

(18) $d = v/f$

With this, all the basic concepts have been introduced, and Euler can now pass on to the explanation of colour phenomena. This is not done in a direct way, but via Newton's theory of colours. For this he must first state the way in which Newton's ideas relate to his own wave theoretical concepts. Euler divides the rays into two kinds, the first being 'simple rays', in which the pulses follow each other at regular time intervals. Euler identifies these simple rays with Newton's 'homogeneous rays', and – as in Newton's theory – the perception of a particular colour is linked to a particular simple ray. Euler believes the colour to be determined by the sole constant property of a simple ray, the frequency of the pulses. He has to render acceptable this idea that is now so obvious to us, and he does so by positing that it will make a difference how often in each second the organ of sight is hit by ether pulses. Moreover, he states, the perception of high and low tones is based on a similar process. Euler's second kind of ray is the 'composite ray', in which the pulses follow each other at irregular time intervals. However, he feels that the name composite ray is less precise, since his theory involves no composition of rays: A 'composite ray' is a single ray produced by a nonisochronic vibration, rather than a mixture or collection of different rays. Even so, he prefers not to deviate from the established terminology derived from Newton.[76] Viewed historically, this is an unfortunate decision, since Euler is obscuring the existence of a significant difference between his ideas and those of Newton, at any rate where Euler's (probably not unjustified) interpretation of Newton's ideas is concerned. It is true that in one passage Euler talks of two kinds of composite rays, but because he is less careful with his terminology in other places, the reader might conclude that Euler and Newton are in agreement. Moreover, in the passage dealt with here, Euler states that Newton's composite ray (a collection of simple rays) has almost the same effect on the senses as a 'really' composite ray. Here he draws attention to the analogous phenomenon in which several well-chosen tones sounded together are heard by the ear as similar to the sound of a single, nonuniformly vibrating chord. Later on Euler, in his treatment of the refraction of white light and the chromatic dispersion occurring with it, places greater emphasis on the ways in which he differs from Newton. He then tries to demonstrate that solar rays are really composite and not of Newton's kind.[77]

c. Reflection and refraction

We noted earlier that Euler in treating refraction and reflection made no use of Huygens' construction involving secondary spherical pulses. As with others before Euler,[78] the treatment of reflection amounts to a reminder of the laws of collision for macroscopic bodies.

He is even prepared to state that, given these laws, reflection is easily explained in both emission and medium hypotheses.[79]

This generosity is more apparent than real, for Euler is now able to skirt round the difficulties created by the naive use – in *both* optical traditions – of the laws of collision. Newton had, after all, seized upon the irregularity of the surface and the phenomenon of internal reflection, for example, in order to suggest that the collisions of light particles with the surface of a medium could not explain the reflection of light. He believed that there had to be a force divided uniformly over this surface to produce reflection.[80] This provided no satisfactory solution to the problem, merely transferring it to the physical explanation for this force. Euler's apparently tolerant attitude towards the emission tradition masked the problems that remained to be solved by the medium tradition on this point as well. Euler reveals his most mathematical side in his discussions of reflection and refraction. It is as though he wants to hurry towards what most interests him, as a mathematician and as a physicist: dispersion and the composition of white light.

First, however, he has to deal with refraction and in doing so Euler makes use of two suppositions. First, he assumes that the pulses in a denser medium are propagated less rapidly than those in a less dense medium. He thus opts for the velocity ratio that was to be regarded as normal for a wave theory from mid–eighteenth century on. He returns later in his treatise to the physical cause for the velocity reduction in a denser medium,[81] but here he passes over this problem in silence. Second, he utilises a proposition that had already been incorporated into his definition of a light ray, that is, that rays of light are perpendicular to the pulse fronts (Euler: pulses). This proposition originated with Huygens and had become common property by the mid–eighteenth century.[82]

Euler now describes the refraction of light (see Figure 9). Pulse *Pp* is located at the transition from medium *ACB* to the denser medium *ADB*. The part of the pulse indicated by *P* moves at velocity v_2 within the medium *ADB*, and that marked *p* remains in medium *ACB*, moving at velocity v_1 until it reaches the surface *AB* at point π. At that moment the pulse part *P* reaches point Π. For the distances covered one has

$$(19) \qquad p\pi \,/\, P\Pi = v_1/v_2$$

Euler indicates that the pulses *Pp* and $\Pi\pi$ cannot be parallel, since v_1 does not equal v_2. He also observes that the rays are perpendicular to

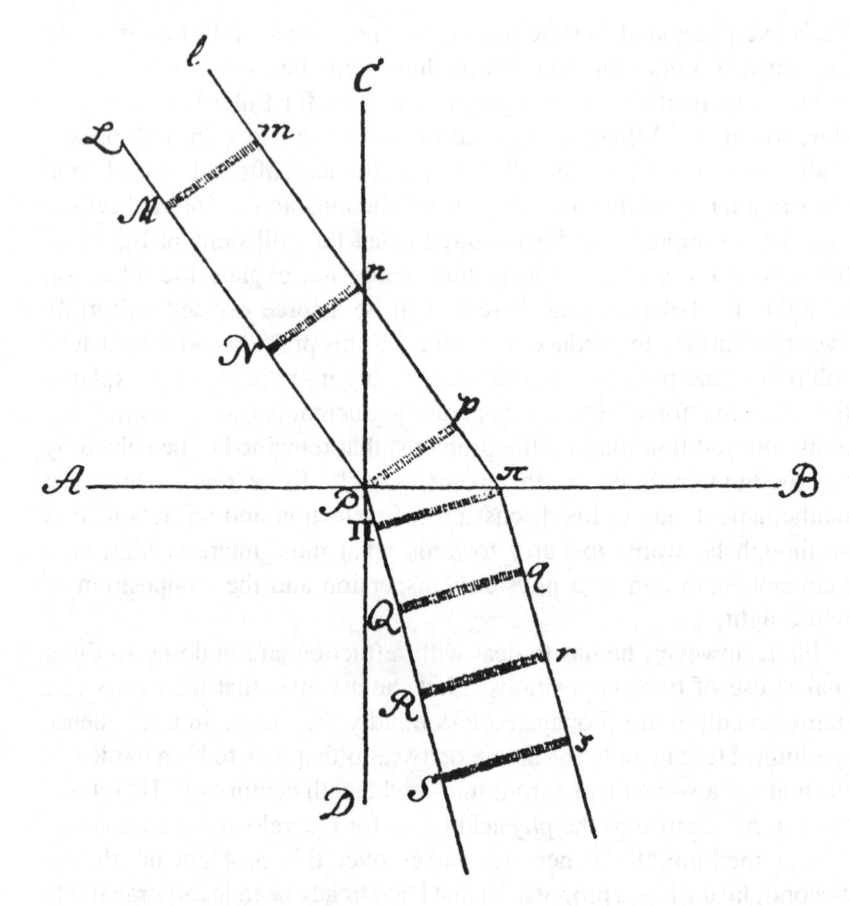

Figure 9. Euler's view of the refraction of light (Euler, 'Nova theoria', 26).

the pulses. Both these data taken together justify the geometry of Euler's refraction diagram.[83]

Euler then deduces, from relation (19), the law of refraction in the form (i = angle LPC, r = angle DPS):

(20) $\sin i/\sin r = v_1/v_2 = \text{constant}$

For Euler one advantage of his deduction of Snel's law is that it respects the principle of least time, which in his view is a most important natural law. Since the deduction of the law of refraction using this principle is generally known, Euler needs only to remind his reader of the main points of this method.[84]

d. Colours: Dispersion and white light

The separation of colours in refraction constitutes a difficulty for emission theorists, as Euler rightly observes. Allotting different refrangibilities to light particles, by assuming a difference in their mass or velocity, for instance, they had not solved all their problems.[85] Euler, however, also believes – wrongly – that his wave theory provides an adequate explanation of refractive dispersion. Nonetheless, his treatment of dispersion possesses great importance for the historian. After Johann II Bernoulli's sketchy hypothesis, it was Euler who produced the first lucid and detailed proposal for a wave theoretical explanation of dispersion.

According to Euler, his theory of light at first sight seems unsuited to the elucidation of dispersion: In the theory the cause of refraction lies in the difference of velocity between two media; consequently dispersion can only occur where there are discrepancies in the velocities of the rays before refraction. Since the red rays are less refracted than the violet, the red rays must have the greatest velocity, for a high velocity in the refracting medium corresponds to a small index of refraction. There is a problem here, arising from the assumption in Euler's theory that pulse velocity is not dependent upon the size of the distance between pulses, but rests exclusively upon the elasticity and density of the medium. Euler had already hinted at the solution earlier on: A uniform velocity only operates if the successive pulses are mutually independent. For the explanation of dispersion he drops his previous supposition and now assumes that the pulses influence one another. It is obvious to Euler that an influence of this kind results in an increased velocity for a pulse sequence, as compared with that of a single pulse. He is uncertain about the mechanism of the influence. The pulses 'act' on one another, he states, and it is 'easy' to recognise that, because of this 'action', the velocity of the pulses is somewhat greater than that of an individual pulse.[86]

Euler subsequently refines this basic idea. Where the difference in velocity between a pulse sequence and that of an individual pulse is concerned, the higher the frequency of the pulses, the greater the difference, whereas the greater their velocity, the smaller the difference. For his calculation Euler chooses the following expression of the velocity difference:

(21) $v - v_0 = \beta\,(f/v_0)$

Here v_0 is the velocity of an individual pulse, β a proportionality constant (positive), and f the frequency. Euler indicates that a first approximation has already been made: He should have used v, and not v_0, in the right-hand term of (21).

If v_i is the velocity of the sequence in medium i ($i = 1,2$), the refraction index n ('refrangibility', in Euler's words) is

$$(22) \quad n = \sin i / \sin r = v_1/v_2$$
$$= \{v_{0,1} + \beta(f/v_{0,1})\} \,/\, \{v_{0,2} + \beta(f/v_{0,2})\}$$

Since the factors $\beta(f/v_{0,1})$ and $\beta(f/v_{0,2})$ are small, it follows from (21) and (22) that

$$(23) \quad n \simeq v_{0,1}/\{v_{0,2} + \beta f\{(v_{0,1}^2 - v_{0,2}^2) \cdot v_{0,1}^{-2} \cdot v_{0,2}^{-1}\}\}$$

The inequality $v_{0,1} > v_{0,2}$ applies to the passage of light from air to glass, so that the second factor in the denominator is positive; that is, the rays undergo less refraction as their frequency increases. Therefore, the way in which colours become separated depends upon the frequency distribution of the incoming rays. In the deduction of the dispersion equation it is manifestly clear that the sole constant factor, and therefore the factor determining the colour of a homogeneous ray of light, is neither pulse velocity nor pulse distance, but the frequency of the ray.[87]

From this dispersion equation Euler deduces that the colour red has the greatest frequency in the solar spectrum, for red is the least refracted, and the refraction index decreases with rising frequency. It subsequently became clear that Euler was wrong and that violet, on the other side of the spectrum, has the greatest frequency; nonetheless, his conclusion and particularly the way he reached it, constitutes an important step forward in the history of the medium tradition. As we have seen, Malebranche despaired of ever being able to determine absolute frequencies, although he claimed that red had the lowest frequency and violet the highest. He based this idea on the vague equation of the 'powerful' colour red with the 'greater' displacement of the concomitant vibrations.[88] Euler introduces an alteration to the problem. He, too, doubted whether absolute frequencies would ever be determined, because of the extreme and therefore 'immeasurable' velocity of pulses. He is more optimistic, however, with regard to ordering the frequencies. In contrast to Malebranche, Euler bases his ideas on a quantitative and in principle testable theory of dispersion rather than on vague analogies. This unambiguous quantification compares favourably with the ambiguity of earlier reasoning, and the difference from what had gone before is even greater when we look at a second consequence of Euler's approach. It is precisely because frequency constitutes the central concept in his wave theory that there are more points at which the hypothesis that red possesses the highest frequency can be put to the test. Euler carried out tests of this kind several times, yet he was unable to find his way onto firm ground. In 1750 he changed his mind on the red/violet question, thinking that violet rather than red had the highest frequency. He retained this idea in an article of 1754, basing it on Newton's data on colours in thin plates. In 1756 he returned to his first opinion that red had the highest frequency. Finally, Euler came to the conclusion that the problem was insoluble at his point of time, since, as he said in the *Lettres* (1768), it was not known whether red or violet had the highest frequency.[89] We would be going too far off the intended track if we examined this fascinating episode in Euler's optics in any greater depth, but we would do well to remember that Euler changed his mind on the red/violet question several times; he was even reproached for this by a German critic. However, the details of his dispersion theory of 1746, or the reasons why he changed his mind at later dates, played no role in this criticism.[90]

It should surprise us if Euler's treatment of the red/violet question were to differ completely from Malebranche's approach, as it would seem to do from the account up to this point. However, historical continuity is also preserved in this case. In his 'Nova theoria' Euler provided, besides his physicomathematical reasoning, a second argument for the proposition that red had the highest frequency, and this belongs in the familiar vaguely expressed genre. Euler compares the colours in an alcohol flame with those in a candle flame. The alcohol flame and the lowest part of the candle flame show a blue colouration, whereas the top part of the candle flame is red. According to Euler, other experiments (which he does not specify) show that the alcohol flame has less power; in addition, the motion at the base of the candle flame is 'without doubt' less than that at the top of the flame. The more marked the motion, the higher the frequency, so that the alcohol flame and the bottom part of the candle flame show the low-frequency blue colouration, and the upper part of the candle flame the high-frequency red.[91]

Although Euler's kind of arguments are partly common in the medium tradition, he overtook his predecessors by proposing a dispersion formula and deducing from it the colour with the highest frequency. Euler was still dissatisfied with this, wanting to know more about the circumstances of dispersion: about the way in which a sun's ray is constituted and the way in which the variously coloured rays are formed from this. We are already familiar with his answer to the first question, having encountered his concept of composite rays and thus of white, solar rays. Euler's idea is supplementary to that of Newton, at any rate as Euler interprets him. According to Euler, Newton regarded a composite light beam as a collection of parallel rays of every colour. In Euler's opinion, however, there is a second kind of composite beam in which differently coloured 'parts' occur in succession within each ray. He posits that solar beams are of the second, 'Eulerian' kind.

What argument does Euler advance for this? He argues that, *within a wave theory*, the formation of 'Newtonian' beams conflicts with the simplicity of Nature. On the surface of the sun the particles producing isochronic vibrations with varying frequencies would have to be distributed evenly, so that bundles of rays emerging from every point in the sun would contain the correct proportion of the various colours. However, because the sun's particles are in continual motion (as a result of the high temperatures), at some point in time one frequency will be overrepresented, and at a later time, another. Consequently the composition of sunlight would not be constant; its colour would continuously change. Euler's conclusion is that solar rays do not consist of 'Newtonian' bundles of rays.[92]

It is as well to bear in mind that Euler's disagreement with Newton on the nature of white sunlight should not be perceived in the context of the debate on the medium and emission traditions. Euler is saying two things. First, he proposes to extend Newton's theory of colours introducing a second kind of composite ray. This suggestion is part of a descriptive theory of colours, with no immediate consequences for an underlying medium or emission conception on the ontological level. Second, he argues that, *in his wave theory*, sunlight can be composite only in the Eulerian sense, an idea that gives indirect support to this first suggestion. Here, too, there is no direct confrontation with the emission tradition. Once again, Euler is obviously distinguishing, as his predecessors in the medium tradition had done, between Newton's descriptive theory of colours and an emission conception often ascribed to Newton. The difference between Euler and Malebranche or Johann II Bernoulli is that Euler adopts Newton's theory of colours only with alterations.

Now that Euler has rejected the idea of 'Newtonian' collections of rays for his own wave theory, only one option remains to him. He

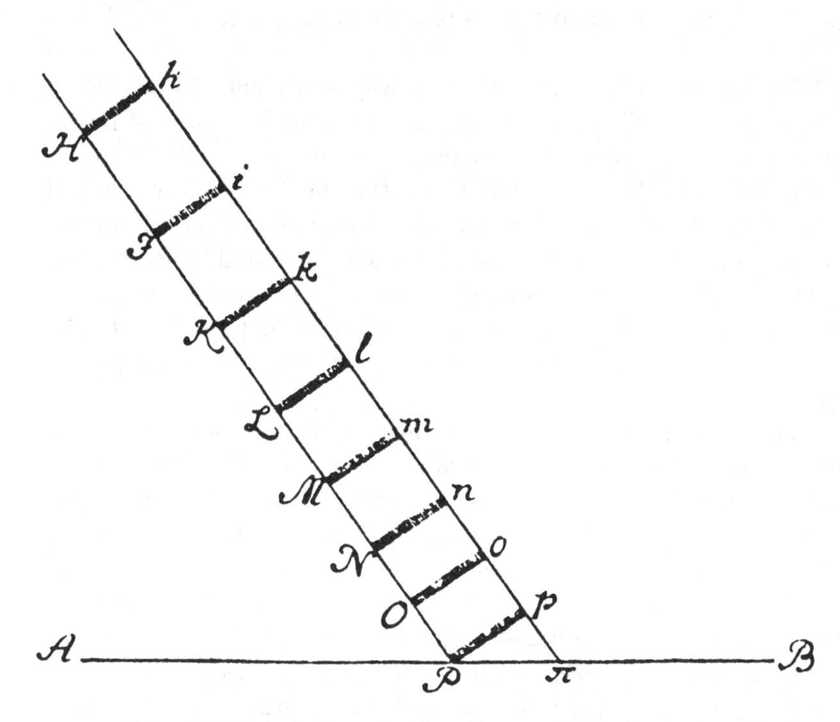

Figure 10. Euler's view of white solar rays (Euler, 'Nova theoria', 33).

conceives a white ray of light as one in which the intervals between the pulses vary in size. He provides a direct argument in support of this conception of white sunlight, in addition to the indirect support discussed above. Motion within all 'fires', and especially in the sun, is extremely powerful. The particles are put into a state of vibration by forceful explosions, so to speak, and therefore cannot be isochronic. Just as the musical instrument string that is struck too hard gives a higher tone at the beginning of its motion than it does at the end, so the pulses in a sun's ray will follow one another more rapidly at the beginning than at the end. The solar particle vibrates with the highest frequency immediately after it has been 'struck'; the successive vibrations gradually decrease in number until the particle is 'struck' again, and so on.[93] In Figure 10 Euler represents the rays formed in this way. It is no coincidence that *HI* in the diagram is almost twice as long as *OP*, or that seven pulse intervals are shown. The last provides an implicit reference to Newton's seven principal colours. The first property is connected with Euler's idea that the frequencies in the solar spectrum constitute almost an octave (a frequency ratio of 2:1; see below).

What happens when the sun's rays shown in Figure 10 fall onto a refracting medium? According to Euler's dispersion rule (the higher the frequency, the less the refraction), the pulses *Pp* and *Oo* are refracted less than the later pulses *Ii* and *Hh*. We could talk in terms of red and violet pulses at the beginning and the end of the rays, respectively, although this is somewhat misleading. It would be more accurate to talk about red and violet pulse *distances* or, even more precisely, about red and violet time intervals between the pulses. (Naturally the meaning is always that of red-*making*, rather than of red pulses, and so on.)

Euler himself raises the problem that, in his representation of the matter, the pulses in the refracted rays do not follow one another immediately. We could say that refracted rays are 'pulse trains' or, even better, arrays of 'pulse coaches', alternating with 'pulseless spaces' measuring the equivalent of several or even a great number of 'coaches'. But how, then, could the refracted rays still form a unity? In Euler's view they do so because the same order of pulses soon recurs within the solar rays *HPhp* and because the empty places are filled by pulses from the rays produced by neighbouring particles in the sun.[94] The first idea accords with Euler's earlier argument that a particle in the sun begins its vibrations, continuing to vibrate ever more slowly until it is once again 'struck'. This means that a new 'pulse train' follows immediately upon pulse *Hh*, the new pulse sequence having the same time intervals as that of 'pulse train' *HPhp*. The second idea – the filling of the 'pulseless spaces' by pulses from neighbouring rays – is not clearly formulated. Euler cannot mean that the pulses spread out sideways, for this would contradict his thesis on the strictly rectilinear propagation of light. We should perhaps read this passage as meaning that the eye is struck not by only one ray, but by a collection of rays, and that the 'pulseless space' in a ray of a particular colour consequently goes unnoticed because of the presence of pulses in neighbouring rays of the same colour.

e. Colours: Opaque bodies and undercolours

In Chapter 5, the conclusion of the 'Nova theoria', Euler discusses the sequence of four kinds of body that can be distinguished in optics: light giving, reflecting, refracting, and opaque objects. The discussion of the fourth kind is the most important. Concealed within Euler's treatment of this subject is the introduction of what we could call undercolours and overcolours, that is, colours with frequencies that are lower or higher than that of sunlight. These special colours and the subject of opaque bodies will be emphasised in this section.

In his treatment of reflecting bodies, Euler claims that the elasticity of these bodies would be achieved more easily by polishing.[95] This shows that, earlier in his treatise, Euler quite rightly remained silent about the physical explanation for the reflection of light. From what he says, it is clear that either he was not sufficiently aware of the impossibility of achieving a microscopically completely smooth surface by polishing or he attempted to circumvent this difficulty.

Where refracting bodies are concerned, Euler undertakes a deeper investigation of the physical properties of the bodies involved. He nevertheless does not pose significant questions or, if he does advance them, he answers them as a ruler, by decree. Euler believes that the pulses within a medium – glass, for instance – are propagated exclusively in the glass particles, rather than in the ether among the glass matter. The glass particles have a density almost infinitely greater than that of the ether particles. If one used equation (11), the standard formula for velocity within a medium, for calculating the velocity of light in glass, the resulting value for the velocity of propagation would be much too low. But Euler thinks it is incorrect to employ equation (11), because propagation in glass is entirely different than propagation in the ether, since the conjunction (*coniunctio*) of the parts of the glass causes the pulses to be propagated as in an instant. (This seems to echo Newton's explanation for the difference between measured and calculated velocity of sound, which, as we know, Euler rejected.) This mechanism ensures that the velocity in glass is much greater than can be expected on the basis of equation (11) and is 'not very different from the velocity in the ether'.[96] This last statement is committed to paper without Euler's drawing attention to the rather coincidental nature of the fact that two entirely different propagation mechanisms result in virtually the same velocities. Neither does he touch upon the fact that various differing materials, such as water, glass, and diamond, demonstrate almost the same propagation velocities.

Rather than proposing solutions to these problems, or at any rate taking notice of them, Euler raises another question connected with the propagation of light within refracting media. If this propagation is rectilinear, the particles must be connected so as to transmit pulses along straight lines. Euler formulates this as a condition for transparent bodies.[97] Doing this, he carries the matter somewhat further than in the velocity issue, at any event stating the problem, even if he offers only a verbal solution.

The largest part of Euler's last chapter concerns the colours of opaque bodies. As always where colours are involved, Newton's theory of colours is the point of departure. However, in the case of opaque bodies the theory is severely criticised. In Newton's theory permanent colours are explained by selective reflection. White rays of light are refracted on the surface of an opaque body, thereupon separating out into rays of various colours. Subsequently only the rays that accord with the colour of the body are reflected, the remainder being absorbed. Here Euler introduces the objection that both reflection and refraction depend upon the angle of incidence, the colour of a body remaining the same regardless of the angle we are observing. Euler explicitly excludes those bodies in which the colours vary according to the angle of vision. He believes these to constitute the exceptions that do not require the same explanation as the majority of coloured bodies, in which colour does not depend upon the angle at which these bodies are viewed.[98]

This observation may be a cryptic reference to Newton's conception, which is the reverse of Euler's. What Euler takes to be exceptions – soap bubbles, the peacock's tail, thin layers of air, as in Newton's rings – are the exemplary cases for Newton. In the latter's view coloured bodies have small transparent particles of a specific thickness that function as a filter: Rays of a given colour are reflected, all the others being allowed to pass on to be absorbed in the end. The variable colours in soap bubbles and peacocks' tails are introduced as confirmation of this explanation. Contrary to what Euler suggests, Newton did in fact recognise the problem that in most bodies colour does not depend upon the visual angle. Newton's solution was to assume that the density of small particles of matter is much greater than that of the surrounding medium. He had noticed on the macroscopic scale that angular dependence decreases where there is a wide difference in density.[99] Euler may have been dissatisfied with this solution, but there may also have been other reasons why he omitted the physical background to Newton's ideas on selective reflection. Whatever the truth of the matter, Euler considered the argument on the angular dependence of colours to be sufficiently powerful to justify the rejection of Newton's explanation of permanent colours.

Since Euler envisages only two possible explanations for the colours of opaque bodies, and since he has repudiated the first (Newton's solution), he is able to push his own ideas to the fore all the more effectively as the only remaining option.[100] Euler believes that a ray falling onto an opaque body is not reflected. On the contrary, it gives rise to a vibration in the surface particles of the body that in turn creates pulses in the ether or in the surrounding transparent medium. The difference between light-giving and opaque bodies is that the first kind is capable of causing vibrations in surface particles, whereas the second kind has to derive this capacity from incident rays of light.

Euler compares the vibrating particles with taut musical instrument strings, each suited to vibrating only at a fixed frequency. A string begins to vibrate, even without being bowed or plucked, if a particular motion in the air is produced by another string that is vibrating at the same or a consonant frequency, that is, by resonance. By analogy, Euler claims, particles on the surface of an opaque body begin to vibrate when rays of light with the same or similar frequencies reach them. We could thus call them resonance particles, the colour of a body being determined by the degree of tension in them. Here, as in the case of light from the sun and other light sources, colour is linked with a particular frequency. In Euler's view the only element still missing from the theory of colours is knowledge of the frequencies corresponding to individual colours. As we have seen, he expressed considerable doubts about the possibility of determining by experiments the absolute frequencies.[101]

It is tempting to claim that Euler modelled his resonance particles on de Mairan's and Johann II Bernoulli's resonating colour particles. After their contribution, it remained for Euler to transfer the colour particles in the ether to the surface of opaque bodies. However, we need to tread cautiously here because it is just as likely that Euler extracted his model from contemporary physicomathematical investigations on cords and strings, combining these with Newton's idea of selection. In the latter case there would be a certain degree of similarity to the origins of de Mairan's resonance particles. De Mairan's inspiration derived from Newton's colour particles in free space, and Euler might have replaced Newton's 'colour-producing particles' in an opaque body with resonance particles in the same medium. Even in the first half of the eighteenth century there were many roads leading to Rome.

After Euler has introduced the basic idea of resonating particles, he begins a more thorough discussion of the different kinds of coloured bodies. If the particles have varying tensions and are spread equally over the surface of the body, the latter has a uniform colour, sometimes only apparently homogeneous; in these cases the heterogeneity becomes obvious in refraction. Thus, a green colour is produced not only by 'green' resonance particles, but also by a combination of 'blue' and 'yellow' ones. A white body requires particles with all tensions mixed evenly. In a black body the particles are too slack to be set in motion; it is the mixing of slack and taut particles that produces the innumerable grades of brightness.[102]

Euler's explanation of the white hue of bodies is not consistent with his earlier objections to the constancy of 'Newtonian' solar rays. After all, there are also high temperatures and heat motions in some white bodies, and in Euler's opinion this renders a constant white colour unlikely. Euler appears to have experienced some difficulty in fitting the idea of white bodies into his theory. In his 'Pensées' he had introduced special particles that vibrate not at a fixed frequency, but at the frequency impressed upon them.[103] There could be no comparison of such particles with taut strings, and it is conceivable that Euler abandoned a separate category of 'white' particles for the sake of simplicity. However, he did not appreciate the fact that he had dismissed his own explanation of the constant white hue of opaque terrestrial bodies as being extremely unlikely for the sun's surface.

In Euler's view the brightness of the colours also depends upon the intensity of the light striking a body. The stronger the original pulses, the more strongly the resonance particles are set in motion, which in turn produces correspondingly stronger pulses. At this point he also discusses the burning and melting of materials by means of focussed mirrors and lenses: The collective force of the pulses breaks the particles of matter into pieces or drives them apart. This idea, according to Euler, advances our knowledge of burning and melting, 'since all other phenomena accompanying these changes in the bodies fit wonderfully well into this'.[104] Euler's lavishly vague formulation refers probably to light and heat effects occurring in melting and burning. This is one of the few places in his 'Nova theoria' in which Euler engages a more chemical subject.

The conclusion to the 'Nova theoria' is devoted to the simultaneous occurrence of the four 'properties' of luminosity, reflection, refraction, and opacity.[105] Here he also provides a brief discussion of the phenomenon of phosphorescence. In sunlight the surface particles of the Bologna stone (a solar phosphorus) acquire a vibratory motion lasting so long that the stone continues to emit light even when there is no incident light. As we shall see in the next chapter, Euler worked out his explanation of phosphorescence in greater detail in response to the results of new experiments. Remarkably enough he barely mentions the joint occurrence of refraction and reflection when he is dealing with mixed forms of the optical 'properties': Euler neglects the topic of partial reflection, like most natural philosophers in both the medium and emission traditions.

To conclude the discussion of Euler's 'Nova theoria', we return to his elucidation of the colours of opaque bodies. Here we find a passage that deviates somewhat from the line of his argument, but that can nonetheless be regarded as the culmination of Euler's wave theory of light. This passage begins at the point where Euler has just written about the linking of colour and frequency. Without any preparation, the reader is suddenly confronted with a startling hypothesis:

> Let us suppose that a ray representing a red colour, carries f pulses to the eye in one second; and, just as in music sounds are held similar which have vibrations, produced in the same length of time, that bear a double, quadruple, eight-fold (etc.) ratio [to the main tone], so simple rays containing, in one second, 2f, 4f, 8f (etc.), or ½f, ¼f, (⅛)f (etc.) vibrations, will all be considered red.[106]

In his 'Pensées' Euler adds: 'comme je prouverai plus bas', but the promised evidence was never produced in this text.[107] What Euler may have had in mind in 1744 was that the idea of undercolours could be applied to the explanation of Newton's rings, yet he gave this explanation only ten years later, in 1754, and never returned to it again. Perhaps between 1744 and 1746 he did not have the opportunity to work out in detail the example of Newton's rings. This could be the reason why he provides no further substantiation in the 'Nova theoria' for his hypothesis on under- and overcolours.

In his 'Nova theoria' Euler does elaborate on the analogy between sound and light. Just as there are innumerable tones in a musical octave, only a few of which are used and have been named, so there are innumerable colours, only a certain proportion being named, a number that differs according to the richness of a given language. Each of the remaining colours is usually called by the name of the named colour closest to it, Euler states. He neglects to mention that the latter idea does not accord with musical usage.

The analogy with the musical octave leads Euler to posit that frequencies in sunlight differ from one another by less than a factor of two. After all, the fastest vibration produces a red colour, whereas the lowest frequency is linked with the colour violet; thus, beyond violet, red would have to be visible again should a vibration in the sunlight occur that was twice as slow as that of red. Euler calls colours outside the solar spectrum derivative colours (*colores derivativi*). I use the terms *undercolours* and *overcolours* because Euler's nomenclature is less informative. In view of the close relationship with the first seven tones of a musical octave, we could call the colours of the sun's spectrum the solar seventh. According to Euler, vibrations faster than those of fire particles producing the solar red do not exist at all; thus, there are only derivative colours with frequencies lower than those of the solar seventh. In our terminology, only undercolours occur in nature.

Undercolours should show a higher degree of refraction than the most refrangible colour in sunlight, violet. Euler thinks it worthwhile to investigate this in an experiment. An even more remarkable suggestion for an experimental testing of his hypothesis escapes Euler's pen when he, somewhat further on, suggests the idea that a body of a particular colour could possess resonance particles with frequencies differing from one another by a factor of exactly two. A body of this kind would show a homogeneous colour at first sight; the diversity of the rays would become perceptible, according to Euler, only in the refraction of the rays. He remains silent on this point in the rest of the 'Nova theoria'. Earlier on he posits in the 'Pensées' that the pure and vivid colours are those of the solar spectrum and equates the undercolours with mixed and less bright colours.[108]

4. The influence and significance of the 'Nova theoria'

Euler's hypothesis on the existence of undercolours may be regarded as the eighteenth-century apotheosis of the introduction of the frequency concept into the medium tradition. The leading idea in this hypothesis, the analogy between sound and light, is carried to its utmost limits, and viewed with the benefit of hindsight, Euler even exceeded the limits, since the hypothesis on undercolours clearly had no lasting value. A later hypothesis that was proved correct can, however, be seen as a conceptual transformation of Euler's hypothesis. In 1802 Young interpreted the recently discovered 'heat rays' as rays with frequencies lower than those of red light. In 1804 he demonstrated experimentally that 'chemical rays' (which cause colour changes in silver salts, for example) have frequencies higher than those of violet.[109] These two kinds of radiation are nowadays known as infrared and ultraviolet. For Young there are thus under- and overfrequencies that, in contrast to Euler's view of the matter, are present in sunlight (but invisible) and do not constitute a repetition of the solar seventh. In spite of these differences, there is sufficient agreement to justify a closer examination of the extent to which Euler might have influenced Young. An influence of this kind would provide support for the hypothesis, recently defended, that Young's wave theory was based upon that of Euler's to a far greater extent than Young himself chose to acknowledge.[110]

Euler was not always wide of the mark; for example, his explanation for the colours of opaque bodies became a permanent element in the medium tradition. This contrasts with his explanation of dispersion, which did not enjoy a long life; it found no audience, and Euler himself made no further use of it after 1746.[111] A last noteworthy element of Euler's wave theory of colours – the Eulerian composite rays and the criticism of 'Newtonian' ray pencils – raises in a lucid way a significant problem in colour theory, that is, the spatial and temporal differentiation of rays or ray parts corresponding to different colours, but without producing a generally accepted solution for it.[112] It is not known to what extent Euler's formulation of this problem influenced later investigators.

Apart from the various hypotheses and explanations connected with the theory of colours, their common foundation, the concept of frequency, is Euler's principal contribution to the medium tradition. The idea of frequency gained its central position through Euler, and it has retained this ever since. Malebranche and Johann II Bernoulli did their best, but it was Euler's work that anchored the concept of frequency firmly in the eighteenth-century medium tradition. Euler's most significant failure was in explaining the rectilinear propagation of light, something in which all his predecessors and every other eighteenth-century natural philosopher also failed.[113]

Euler nevertheless succeeded in creating a full-fledged alternative to the emission theories, thereby fulfilling a necessary condition for having a full-fledged discussion on the nature of light. Only after Euler were there two opposing traditions of equal value, and Euler, together with Newton, initially provided the main issues in the debate. It was tragic for Euler that the discussion revolved almost exclusively round the points of contention dealt with in his introductory first chapter. The details of his wave theory, which had made the discussion possible in the first place, added little or nothing to it. Further, development of the content of Euler's wave theory came to a virtual halt in the half-century following the 'Nova theoria'. Euler's first chapter was written to ensure that the next four chapters would be read, but in fact the existence of the last four chapters ensured that only the first chapter would be used. The 'Nova theoria' fulfilled a historically important function that was, however, of only minimal interest to Euler himself. His treatise on light created the opportunity for a lively debate in Germany because it helped the medium tradition to prevail there. As we shall see, that debate remained undecided for several decades. It is as though history showed compassion in sparing Euler the conclusion of the debate, which would have disappointed him. Several years after Euler's death there was a sudden reorientation in the discussion and a relatively rapid outcome in favour of the emission tradition. Euler died in 1783; six years after this the chemical effects of light, recently discovered, were given a prominent place in the discussion, and a little more than ten years after his death, the matter had been decided against his ideas.

5

The debate in Germany on the nature of light, 1740–95

1. Introduction

In the standard historiography of science the eighteenth century is the period in which the emission conception of light was quite generally accepted, certainly after 1740. Euler is usually mentioned as the exception to this rule.[1] Surveys that are more oriented towards Germany add one or another dissident to the list but leave the image unaltered on the whole: The emission tradition ruled the physical optics roost in the eighteenth century.[2] Apparently the picture of the situation in different countries is to a great extent determined by simply declaring the general picture to be valid for every country, without any thorough investigation of the matter. Britain and Ireland are the only countries on which detailed and systematic research has been carried out. G. N. Cantor has provided an exhaustive survey of optical viewpoints in this region. His results do, it is true, lend nuances to the established image, but they introduce no radical change. In Cantor's book only 9 per cent out of a total of sixty-nine optical theorists from the eighteenth century support a medium theory, while the remaining 91 per cent can be located within the emission tradition.[3] In other words, the historical evidence thus far available confirms the strongly dominant position of the emission tradition in the eighteenth century. Nevertheless, it will be argued in this section that a substantially different view of the matter ought to be given for Germany.

We do not need to seek very far for the outlines of the new picture. To begin with, we have the guidance of a generally reliable source dating from the end of the eighteenth century, Johann Samuel Traugott Gehler's *Physikalisches Wörterbuch*. Gehler stated in 1789 that Euler had developed the medium hypothesis of light so well 'dass man es noch zur Zeit nicht wagen kan, zwischen seiner Theorie und dem

Emanationssystem völlig zu entscheiden'. Although Gehler himself inclined towards an emission theory, officially he adopted a neutral position when positing 'dass beyde Systeme zwar viel erklären, beyde aber auch grosse Schwierigkeiten gegen sich haben'.[4] This indicates that the two optical traditions balanced each other in approximately 1790. Johann Carl Fischer suggested something even more surprising in 1803, in his *Geschichte der Physik*, stating that Euler's theory of light 'von den meisten und berühmtesten Naturforschern eine geraume Zeit hindurch vertheidigt wurde'.[5] If Fischer is to be believed, Euler's theory was indeed considered passé at the beginning of the nineteenth century, but it had enjoyed a strongly dominant position for a long period before that.

The support afforded by general statements such as those made by Gehler and Fischer is not sufficient for any radical corrective to the general interpretation of optics in the eighteenth century. An extensive study of historical sources shows that the truth of the matter lies somewhere between the two suggestions cited. In Germany there was for a long time more than just a balance between the two optical traditions; however, the preponderance of the medium tradition was too slight to permit of talking in terms of a *strongly* dominant position.

The basis for this conclusion is a selective survey of eighteenth-century conceptions in Germany on the nature of light, as represented in Table 3.[6] The years 1716 and 1795 have been taken as the time limits. Since ideas in the second half of the eighteenth century diverge considerably from the historical picture currently accepted, a larger number of sources has been selected from that period than for the time between 1716 and 1745. Mainly printed sources have been used.[7] They comprise physics texts (the majority), research publications, and books of an encyclopaedic kind. Most of the sources have been located by using bibliographies that display no preference for either of the optical traditions.[8] Only the first relevant text by a person is mentioned in the table, unless he or she changed his opinion at a later date or introduced considerable changes into the argument supporting a particular point of view.[9] For these reasons, several people appear more than once, up to three times in one case.[10] The year cited in Table 3 is the year of publication and/or the year in which the text was written or completed, in cases where it was not published in the period before 1795.

The general discussion of the data in Table 3 is divided into two parts. First, we shall examine the conceptions of the nature of light current in Germany in approximately 1740, shortly before the publication of Euler's 'Nova theoria'. This constitutes the point of departure for a more detailed investigation of optical opinions in the second half of the eighteenth century.

a. The situation in approximately 1740

Between the beginning of the eighteenth century and the publication of Euler's wave theory virtually no original contributions to the theory of light and colours appeared in Germany.[11] Consequently the local background to the reception of Euler's theory was to a large extent determined by textbooks and one or two articles in an encyclopaedia. The level of this literature is not always particularly high. This is expressed, for example, in a lack of comprehensiveness and depth often resulting from a form of publication aimed at a general public. It was a handicap for the medium tradition in particular that the mathematical parts of this tradition fell between two stools. This applied especially to Huygens' theory of light, which was too mathematical for a physics textbook and too physical – or more precisely, too natural philosophical – for the optical chapters in a textbook on the mathematical sciences. The fact that the level was generally low did not mean that the important developments did not on the whole reach the learned world. As we have seen, Newton's new theory of colours was absorbed relatively quickly and without problems by German natural philosophers, mainly through the activities of representatives of the medium tradition, ironically. Where the nature of light is concerned, it will become clear that the emission and medium traditions balanced each other in the Germany of the 1740s.

The general features of the German situation are clearly visible in two unsigned survey articles appearing in 1735 and 1738 in Zedler's *Universal-lexicon aller Wissenschaften und Künste*. The first and most comprehensive article treats of colours. Following a concise historical outline, most of the space in this work is devoted to an exposition of Newton's theory of colours, which is presented as the only correct theory, confirmed by observation and experiment.[12] There is no clear choice for any particular conception in the second article, dealing with light. First of all, 'the' properties of light are enumerated: its rectilinear propagation, reflection, refraction, inflection, and finite velocity. Next come the hypotheses that try to explain these empirical data, with discussion of Aristotle, Descartes, Gassendi, Huygens, and Newton.

Table 3. *Conceptions in Germany of the nature of light, 1716–95 (see note 6)*

Year	Emission hypothesis		No choice		Medium hypothesis	
1716 up to and incl. 1745	1727	G. E. Hamberger			1718	Kaschube
	1740	Segner			1723	Wolff
	1742	Hollmann	1738	*Grosses Lexicon*	1738	Winkler
1746					1746	Euler
	1749	Eberhard				
	1750	Krüger				
	1751	Zieglerin				
	1753	Gordon				
		Dalham			1753	Mangold
	1754	G. W. Krafft	1754	Segner		
1755			1755	Eberhard	1755	Kant
					1757	Darjes
					1760	Lambert
					1761	Hube
1765						
	1767	Westfeld			1767	Maler

	1770 Grant		
		1772 Schmid	
			1773 Ludwig
		1774 Béguelin	
1775			1775 Boeckmann
		1776 Klügel	
			1777 Erxleben
	1780 Gabler		1780 Karsten
			1781 A. A. Hamberger
			1782 Suckow
1785			
			1787 Eberhard
	1788 Gren		1788 Werner
	1788/9 Lichtenberg		
	1789 Arbuthnot	1789 Hobert	
	Heinrich	Gehler	
	1790 Gren		
	1791 Lichtenberg		
	1792 Kries		
			1793 Voight
1795	1795 Gehler		

Malebranche is conspicuous by his absence. Huygens' and Newton's hypotheses are considered the best to date. The theories are dealt with in a superficial way; for example, for the explanation of refraction and reflection in Huygens' theory, the reader is referred to the *Traité*, and for rectilinear propagation there is no reference to Huygens' treatment. Newton's conception is treated in a similarly superficial way. In spite of this, the article concludes that both Huygens and Newton explain the properties of light so well that there is difficulty in choosing between the hypotheses.[13] Nevertheless, as we shall see, in general this choice actually was made in Germany. The fact that both emission and medium traditions were equally represented in actual choices is reflected in the cautious conclusion to the encyclopaedia article.

The balance that one suspects existed, on the basis of the article in Zedler's encyclopaedia, is confirmed when we consult two textbooks that were used extensively. We encounter a medium conception in Wolff's textbook on theoretical physics, first published in 1723 and reprinted four times by 1746. In contrast, we find an emission conception in the textbook by Georg Erhard Hamberger, a professor at Jena, that appeared in 1727 and was reprinted for the fifth time in 1761.[14] Other textbooks embodying a medium conception bear the names of Kaschube (1718) mentioned above, and Johann Heinrich Winkler (1738). An emission conception was defended by Segner (1746), among others.[15] As the terminology used here suggests ('conceptions', rather than 'theories'), the chapters on light in the German textbooks do not have the depth and comprehensiveness found in some of the texts treated in Chapter 3. In order to provide an impression of the way in which the nature of light is dealt with, we shall examine the works by Wolff and Hamberger in greater detail.

Wolff's *Naturlehre* embraces a larger area than our present-day physics. His textbook begins with an exposition on bodies in general; this is followed by a section on the system of the world in which the fundamental properties of light are treated, in the chapter on the sun. Next follows a section on the earth, in which the rainbow is discussed, among other topics. The book closes with a section on plants, animals, and humankind.

Wolff believes that the sun sets the all-pervasive subtle matter in motion with its flames. This motion is propagated at a great velocity.[16] Wolff's arguments in support of this medium conception resembles, though in reverse image, a reasoning we have already encountered in the emission tradition: If light is not emitted matter, it is *therefore* a motion within a medium. Wolff advances three arguments. In his view the extraordinarily high velocity of light conflicts with the idea of emitted matter. Moreover, according to an emission conception, a room should not immediately become dark when the shutters are closed, since the light particles are still in the room and ought to be perceivable for a while. Finally, he argues against the emission conception of light, positing that there must be, in general, a gradual increase in the quantity of matter in the places to which a fluid flows. Here we have to imagine perhaps a bucket slowly filling with water and, by analogy, the accumulation of light matter. Whatever the case, Wolff contrasts the behaviour of a fluid with that of light, which does not increase in intensity with the passing of time.[17]

The first argument against an emission conception makes one think immediately of Huygens, whose influence subsequently becomes even clearer. The *Traité de la lumiere* provided the model for Wolff's treatment of light. In the explanation for the velocity of light and the unimpeded crossing of rays, the familiar arrangements of aligned globules resurface one after the other. The elasticity of the ether particles is traced back to the motion of an even more subtle matter; for Huygens this was still a tentative hypothesis, but Wolff advanced it without reservations.[18] Wolff cites the *Traité* for further details on propagation, reflection, and refraction. He says that he is omitting these subjects because they assume too great a knowledge of geometry.[19]

So, Huygens' theory is not really used in Wolff's book on physics. Neither was it applied in his textbook on mathematical, that is, geometrical, optics. As was the custom in mathematical textbooks, in his own book on geometrical optics Wolff gives a treatment of light rays that is neutral with regard to physical theory or physical optics; consequently Huygens' theory on secondary pulses and pulse fronts is not mentioned.[20]

Wolff does not engage the problem of rectilinear propagation. Neither did Winkler in 1738, but Kaschube did mention the problem in 1718. Kaschube disagreed with Newton's argument against the medium conceptions and directed readers to the relevant passages in Huygens' *Traité* and *Discours* for a counterargument, although without examining the content of either Huygens' theory or his answer to Newton.[21]

Kaschube also omitted Huygens' principle. Thus, the factor distinguishing Huygens from other theorists in the medium tradition disappeared, the result being a paradox. Huygens' theory was more often taken up by German textbooks than by the specialist literature dealt with in Chapter 3, but in such a superficial way that the essence and power of his theory – Huygens' principle and its concomitant mathematics – were lost.[22] The acceptance of Huygens' theory in such a manner meant that it was all the more effectively forgotten, and a dual verdict on Huygens' theory was therefore reached. It was rejected in the specialist literature of the first half of the eighteenth century, while in the textbooks – in the German ones, at any rate – it fell between the two stools of physics and mathematics.

As in all the textbooks mentioned above, Newton's theory of colours is also included in Wolff's work at the descriptive level. However, its medium theoretical basis is superficial and inconsistent. Wolff assumes that coloured rays are distinguishable by their various 'velocities', rather than by the various 'quantities of subtle matter moved'. He believes these 'quantities of matter' to be responsible for the intensity of light. 'Velocities' here appears to refer to propagation velocities, at all events not to frequencies, about which he is completely silent. He posits implicitly that in the succession white, yellow, red, blue, and black, white light has the greatest 'velocity'.[23] This is at odds with the proposition in Newton's theory of colours that white light is composed of the colours of the rainbow. This obvious contradiction appears to have escaped Wolff's notice.

A feature of the medium tradition in Germany in the period following Euler's time that attracts our attention originates in the first half of the eighteenth century (see Section b, below). It concerns the explanation of refraction, particularly in textbooks. As far as we know, Winkler was the first, in 1738, to hold a conception that has up to now escaped the notice of historians. Following in the footsteps of the 'Newtonians' within the competing emission tradition, the medium theorist Winkler traced refraction back to an attractive force exerted on rays of light by the denser medium. The use of forces within the medium tradition was not excluded, but it seemed to contradict the priority of contact action in the medium tradition (see Section b, below). Winkler considered the existence of an attractive force to be proved by the experiments on the inflection of light. In 1738 Winkler even appears to entertain a physical interpretation of this force, but nearly twenty years later he states explicitly that he regards the attractive force as a mathematical quantity that does not necessarily represent reality.[24]

It is no simple matter to indicate the way in which the 'Newtonian' attractive force was introduced into the German medium tradition. As Huygens' theory was not adopted, or at any rate not in its entirety, the problem how to explain refraction remained to be solved. The 'Newtonian' refractive force was a likely solution. At the somewhat superficial level of a textbook it does not appear to conflict with a medium conception. Furthermore, the attractive force, in contrast to Huygens' geometrical pulse front constructions, for instance, satisfied the specific demands made by a textbook: It was a simple concept that could be based on a clear experiment. In a period in which the importance of the experiment was acknowledged in every natural philosophical camp, certainly in textbooks, the 'Newtonian' explanation of refraction, linked with an experiment on inflection, possessed an extra attraction.

The emission conception as given in Hamberger's textbook of 1727 has greater depth and scope in comparison with the German medium tradition. Nevertheless, Hamberger lags behind 's Gravesande's and van Musschenbroek's 'Newtonian' systematization on both points. This may be connected with the fact that Hamberger followed the *Opticks* in a less meticulous fashion than his Dutch colleagues. For the rest, Hamberger cites both 's Gravesande's textbook of 1720–1 and van Musschenbroek's *Epitome* of 1726, his citation practice leading one to assume that he also knew Newton's work at first hand.[25]

Generally speaking Hamberger takes van Musschenbroek's line more than 's Gravesande's, which does not necessarily imply a direct influence. Hamberger believes light to be emitted fire, moving rectilinearly. Fire is a very subtle, weakly cohering substance in which the particles are round and possess weight.[26] Any arguments in favour of the emission of light or against a medium conception are entirely absent. Cursory argument, or none at all, seems to be typical of the emission tradition in Germany up to Euler.[27]

Hamberger attributes refraction of light to an attraction exerted by the denser medium, and he appears to accept this force as action at a distance and as the ultimate explanation.[28] Conspicuous by its absence is a discussion of internal reflection, which both Newton and the 'Newtonians' also ascribe to the action of an attractive force; ordinary reflection is mentioned, but a repulsive force is not connected with it. A remark of Hamberger's can even be interpreted as a reference to the laws of collision, although Newton explicitly rejects an explanation involving collision on the surface of objects.[29]

Newton's theory of colours is accepted as a descriptive theory. On the explanatory level Hamberger posits, in accordance with Newton's queries, that a homogeneous ray contains particles of the same kind and that the particles in a heterogeneous ray are a mixture of different kinds. He does not go so far as to specify the differences as discrepancies in size, as Newton did.[30] There is absolutely no trace of Newton's rings, the 'fits', or the ether that explains them.

In summary, it can be stated that there seems to have been about equal support for the two main conceptions of light in the German textbooks shortly before Euler appeared on the scene. German supporters of the medium tradition ostensibly appeal to Huygens, while in reality presenting a vague medium hypothesis that can be linked with nobody in particular. The emission tradition in Germany is connected with Newton, but less strictly and comprehensively than was the case with the Dutch 'Newtonians'. The contrast between the two traditions is certainly there, but without the point being driven home; certain subjects, such as the explanation of inflection and refraction, could even break through the borders between these traditions.

b. Survey of the debate, 1740–95

With one exception[31] the participants in the German discussion viewed the question of the nature of light as a choice between two, and no more than two, traditions or theories (see Table 3). By far the largest group adopted an emission or medium hypothesis (columns 1 and 3); the remainder chose neither, but regarded the two kinds of hypotheses as the only possibilities and mutually exclusive (column 2).[32]

No trace of Euler's influence is to be found among the advocates of an emission conception until 1754, eight years after the publication of the 'Nova theoria'. In four out of the five publications between 1746 and 1754 (see column 1), the alternative medium tradition is indeed discussed and rejected, but neither the theory nor even the name of the Berlin academician is mentioned. The argument against a medium conception revolves almost exclusively round the conviction that a hypothesis of this kind cannot explain the rectilinear propagation of light.[33] Georg Wolfgang Krafft, a former colleague of Euler's from their Petersburg days, was the first to devote attention to the 'Nova theoria'. Krafft discusses at length Euler's first chapter on the arguments for and against the two optical traditions, concluding that Euler's argument is unconvincing, but he does not deal with the details of Euler's wave theory (see Section 2, below). From 1754 on Euler is invariably mentioned (sometimes in combination with Descartes, Malebranche, Johann II Bernoulli, or Huygens), even though up to 1789 the rebuttal of the medium tradition is either absent or brief (the argument of rectilinear propagation). Euler's opponents seldom stop to criticise the details of his theory.[34] The rejection of Euler's theory is a repudiation of the eighteenth-century medium tradition as a whole, not merely of Euler's specific wave theory.

From 1789 there is a radical change in this pattern, and in two respects. First, a lot more attention is given to rebutting the medium tradition. We see this happen in a long letter, in two different answers to a prize question for a competition, and in a German translation and adaptation of Euler's *Lettres*.[35] At the same time the chief argument against the medium tradition changes. Where this had concerned the physical phenomenon of rectilinear propagation, the argument is now advanced, in an emphatic manner, that a medium theory cannot explain all kinds of photochemical effects. (The first and most extensive statement of 'the chemical argument' is treated in Section 5, below.) This chemical argument now regularly appeared, in addition to the usual physical objections, in the short discussions of the medium tradition. Moreover, the argument was considered very important, if not actually decisive (see Section 6, below). An element of continuity in the period before and after 1789 is the way in which, generally speaking, Euler's wave theory was discussed by its opponents. The only feature to be discussed and rejected was the one shared with the entire eighteenth-century medium tradition: the hypothesis that light is a motion within a subtle medium.

The first advocate of a medium theory after 1746 in Table 3 is Josef Mangold. He explicitly adopted most of the arguments from Euler's 'Nova theoria' but added his own to this (see Section 2, below). The fourteen remaining supporters of a medium theory after 1746 usually mentioned Euler's name, too. Again, as in Mangold's case, where there is a somewhat detailed argumentation for a medium theory and against the emission theories, this is at least partly based on the first chapter in the 'Nova theoria'.[36] Only one of the adherents makes a fairly intensive effort to support Euler's viewpoint on the rectilinear propagation of sound and to support, in this way, the latter's retort to the main objection to the medium tradition.[37]

There is a remarkable development where the content of the medium tradition is concerned. Several authors do not adopt Euler's theory of refraction and dispersion,[38] following instead the path already trodden by Winkler in 1738. Refraction and dispersion are traced back to the action of an attractive force exerted by the denser medium. This constitutes a strange addition to a medium theory, one against which Euler himself protested strongly, in a letter.[39] The tension between an attractive force and the basic concepts of a medium theory is probably the reason why this variant disappeared after a while. Whatever the cause, Euler's explanation of refraction and dispersion was obviously regarded as unsatisfactory by some of his allies. Since the differences in content from Euler's theory, which are in themselves worthy of further investigation, had no influence on the German discussion concerning the nature of light, I shall not enter into any deeper examination of this here.

If we consider Table 3 as a whole and ask ourselves about the overall picture emerging from it, we may find it worthwhile, first, to ponder a little on the nature of the conclusions that can be based on a table of this kind. The selective character of the survey, the relatively small numbers, the always rather arbitrary way in which the data are presented, the fact that we count people's opinions without asking ourselves what the significance of their contribution is, all of this points in the one direction: Only general trends can be inferred from the table. This is also expressed in the terminology utilised in cases of this kind: 'balance', 'a majority', 'strong dominance', and so forth. On the basis of the nature of the phenomenon 'support for different theories' (which is, by definition, vague), only a rough-and-ready scale can be utilised; it can be defined as follows. In the case of a choice between two possibilities A and B, 80 to 100 per cent support for A is defined as A's strong dominance, 60 to 80 per cent support for A is a dominance by A, and 40 to 60 per cent support is a situation in which A and B are roughly balancing each other. (Here the assumption is that the group of people who opt for neither A nor B is so small that it can be ignored.)

What, then, were the trends, in the period 1740–95, in opinions on the nature of light? As explained in the previous section, there was a balance between the two optical traditions in approximately 1740, a situation continuing until roughly 1755. From that time on, the medium tradition predominated, and it was to do so for more than thirty years. This state of affairs came to an abrupt end in the years following 1789: The dominance by the medium tradition soon gave way to a strong dominance by the emission tradition.

In comparison with the current image of eighteenth-century optics, the German trends produced the following outline. In about the middle of the century Germany was more or less in step with the rest of Europe, even though there was a balance between the medium tradition and the emission tradition, in contrast to the situation in Britain and Ireland. After 1755 the situation in Germany changed, an alteration that was in diametric opposition to what obtained in other countries, where at that time the emission tradition was strongly dominant. Germany only joined the European ranks shortly before the end of the eighteenth century because there, too, the emission tradition had gained the ascendance, within a remarkably short time.

2. Early reactions to Euler's argumentation

Euler begins his 'Nova theoria' with a general attack on the standard arguments against a medium hypothesis, and by enumerating the difficulties attending an emission tradition. Only after this does he set forth his own specific medium theory, a wave theory, and try to gain acceptance for the idea that this theory describes and explains all optical phenomena. It looks as though Euler could have stopped after his first, introductory chapter. Most of his opponents were content with discussing Euler's argumentation or, more frequently, restricted their attention to repeating the argument about rectilinear propagation, the kind of criticism current before Euler's time. Only occasionally did an opponent devote any attention to the details of Euler's theory. We shall examine more closely two early reactions to Euler's argumentation, reactions that are striking in their extensiveness. The first derives from an opponent of Euler's theory; the second from a supporter.

The critical discussion of Euler's arguments was written by Georg Wolfgang Krafft and can be found in his physics textbook of 1754. After a concise historical survey of the various conceptions of light, Krafft devotes ten pages to a detailed examination of Euler's reasoning. He goes through the arguments one by one, always arriving at the conclusion that the author of the 'Nova theoria' is wrong.[40]

In his first argument Euler posits that, where sound is concerned, nature does not utilise effluvia as it does in the case of smell, because of the considerable distances involved. Rather, it makes use of the propagation of pulses in the air. Euler believed that, in order to bridge the much greater distances involved in seeing, light also has to propagate by means of pulses in a medium. Krafft's objection to this is that analogies 'are too metaphysical'. He does not allow the matter to rest with this general remark, but shows in detail how Euler's argument from analogy can be refuted. In answer to Euler's reasoning he introduces the idea that scents and sounds can both leap more or less the same distances, basing this on observations of the distance at which scents emanating from the island of Ceylon can still be detected, and the distance over which a cannon can be heard. With this, Euler's basic argument is countered, but Krafft goes still further. On the basis of the analogy Euler had drawn and the observations Krafft had mentioned, it would be possible to conclude, erroneously, that both sounds and scents are spread by way of effluvia. Krafft says that such faulty conclusions result from placing too much trust in reasoning by analogy.[41]

Euler's second argument concerns the vacuum. Euler, defending a medium theory, maintains that the adherents of an emission theory were unable to sustain their claim that the universe is an entirely empty space, since according to their own beliefs there is at least emitted light matter. To begin with, Krafft posits – correctly – that Newton had not adopted the idea of an absolute vacuum and that the density of the rays decreases in inverse proportion to the square of the distance to the sun. Krafft subsequently explains that there is nothing absurd in the assumption that a stream of light particles goes out from every point in a luminous object to every point in space, provided one assumes that the particles are extremely small and move extremely fast. Consequently there is no question of an absolute vacuum, but space is assuredly not entirely filled with light matter. Krafft concludes that this solves the problem of the alleged contradiction.[42]

Euler also argues that light particles not only fill space but also move at a great velocity – another reason for light matter's impeding the planets in their motions. Krafft conceives this as a separate argument against an emission hypothesis, whereas it was only intended to clear the way for a medium theory. Is attack, after all, the best form of defense? At all events, Euler was outstandingly successful in diverting Krafft's attention from the problems of a medium theory attached to the introduction of an ether distributed throughout space. In his response Krafft disagrees with Euler's suggestion that light matter, moving at enormous velocity, has a noticeable influence on a body as large and heavy as a planet. True, the particles do move extremely rapidly, but the density of light matter diminishes with increasing distance from the sun, and the particles are unimaginably tiny. Krafft makes no calculations and, like Euler, he makes no use of the figures in Segner's *De raritate luminis* or of van Musschenbroek's estimate of the subtlety of light.[43]

In his historical survey Krafft speaks of Newton's thesis on the spreading of waves and pulses into the geometric shadow of an opening. An important link in Euler's defense against this weighty argument is his assurance that nobody in a room with an open door claims that the sound comes from the direction of that door. Krafft coolly remarks that Euler contributes nothing against Newton's theoretical deduction and, as far as experience is concerned, that he has himself frequently observed that sound heard in a room came from the direction of the door opening.[44]

The last two of Euler's arguments dealt with by Krafft constitute explicit objections to an emission theory. The first concerns the loss of matter by the sun. To begin with, Krafft refers to the circulation of light matter: Just as stones thrown high into the air fall back to the ground, so light particles return to the sun after some time. He also reminds the reader that solar rays are extremely subtle. Euler argues, against the latter idea, that an absurdly large degree of fineness in the rays of light would be required to ensure that the shrinking of the sun would not be noticeable. Basing his calculations on certain assumptions, Krafft calculates that a light particle is forty-three times as small as a particle of copper, which in his view shows that there is nothing absurd in the high figure Euler gives to the subtlety of the light particles.[45]

Euler's second reservation about an emission theory concerns the crossing of light rays. If light consists of emitted particles, these must disturb each other's motion. However, no trace of such a disturbance is observed, either in a small opening or in the focus of a lens or mirror. Therefore, light does not consist of emitted particles, according to Euler. Krafft does not seem to know what to do with this argument, or else he has only half his mind on the subject, since his answer to this problem is extremely weak. He admits that the rays will impede one another in the cases mentioned, indeed weakening the light, but, he claims, obviously a large proportion of the rays is not obstructed: 'What other explanation can there be for the fact that the sun can be seen even through the tiniest hole?' Krafft asks rhetorically. He adds: 'What is the explanation for the fact that, just outside the focus of a mirror, a substance no longer ignites immediately?' Here Krafft is making an elementary logical error in assuming the point he wants to prove, that is, that an emission theory is correct, and deduces from this that the mutual disturbance among particles is 'obviously' not sufficient to produce noticeable results.[46]

Curiously enough Krafft, who up to this point has dealt with all of Euler's arguments in their original order, does not discuss the last of Euler's objections, that concerning the structure of matter. In his conclusion Krafft does examine Euler's remark that the best support for a theory resides in its harmony with the phenomena. Using Euler against Euler, Krafft refers to the former's article on the aberration of light in which this phenomenon is explained on the basis of an emission theory. Krafft consequently ends his discussion of Euler's argumentation with the remark that a second opinion is not always the more sensible one.[47] Most of Krafft's readers appear to have noticed this to be a rhetorical turn of phrase, based upon an incorrect interpretation of Euler's article on aberration; virtually none of these readers used the contradiction – more apparent than real – as an argument against Euler's viewpoint.[48]

Both opponents and supporters of a medium theory take Euler's pro/contra reasoning as their point of departure. Josef Mangold, professor at Ingolstadt, is one of the latter. His *Systema luminis et colorum*, published in 1753, is a wide-ranging treatise on light and colours, preceded by a disquisition on sound. Although the size of the section on physical optics, over two hundred pages, arouses the expectation of an exhaustive treatment of the subject, Mangold discusses only those phenomena whose explanation in terms of the nature of light is controversial.[49] In his view these phenomena are transparency, reflection, refraction, and dispersion. He deals with these almost exclusively in a qualitative way. The laws of reflection and refraction, for instance, have no significance in his account. He is concerned with the causes of reflection and refraction, and mathematical laws are clearly not very important in that area.

Within this fundamental and qualitative context he discusses the 'theories' of Gassendi and Newton, Descartes, the Aristotelians, and 'the majority of the moderns'. The latter believe that light consists of a vibratory motion within an extremely subtle, extremely fluid, all-pervasive matter: the ether.[50] He mentions no names here, but where he does so elsewhere he talks exclusively of Euler.[51] Even though Mangold declares himself to be in favour of a vibration theory[52] he is obviously not a slavish supporter of Euler. He diverges from his great example in both the reasoning used and in the details of the theory. He takes his own way even in the manner (described above) in which he introduces the problem of the nature of light. In contrast to Euler, the discussion about light is, for Mangold, not narrowed down to a two-way contest. Moreover, by distinguishing between Descartes' pressure conception and Euler's wave theory, Mangold emphasises the considerable differences between these two variations within the medium tradition. Considering Mangold's environment, it is understandable why he devoted attention to the Aristotelians: In Ingolstadt mechanistic science had only recently gained influence.[53]

If we also regard the argument on the inconceivably great velocity of light particles as stemming from Euler himself, then Mangold is using six out of the eight arguments appearing in the 'Nova theoria'. Besides the enormous velocity he also mentions the shrinking of the sun and the impediment of planetary motion, but he leaves out the two arguments on the crossing of the rays and the structure of matter. One of his other arguments is that, in an emission theory, it must continue to be light for a while in a room where the openings have been closed, just as scents continue to be smelt for some time.[54]

In favour of a wave theory, Mangold first advances the idea that a particular theory is more likely to be true if, on the one hand, it explains phenomena in a satisfactory manner, while, on the other hand, it assumes a mode of action analogous to another, comparable phenomenon, in this case the phenomenon of sound.[55] Euler had also mentioned both these criteria. There is no reference to filled space in Mangold's treatment of the objections to a wave theory. We have just seen that he did adopt Euler's counterattack – emitted light particles also fill space and impede planetary motion – as an objection to an emission theory. Rectilinear propagation is discussed with direct reference to Euler. In Mangold's view as well, sound moves in rectilinear fashion, and walls that allow sound to pass are comparable to windows that transmit light. He obviously feels the need to clarify Euler's statements on this aspect. According to Mangold, air and ether particles are mainly compressed in the direction of propagation. There is little lateral distortion in the case of sound, and even less in the case of light, so, outside the direction of propagation, no perception – or at any rate, only a very weak perception – can be produced in the eye.[56]

After this Mangold discusses another four objections to a wave theory that are not found in Euler's work. The first is formulated in a rather obscure way, but it appears to amount to the difficulty that, by analogy with the loss of sound in the wind, there can be no rectilinear propagation of light in the presence of an 'ether wind'. In answer to this Mangold states that the great velocity of the light vibrations cancels out this effect.[57] The two following objections do not appear to be very serious ones, the intention being rather to create an opportunity for further elucidation of the theory. The first of the two objections is as follows. Since we see stars at the moment that they rise above the horizon, one might think that light is propagated instantaneously; however, an infinitely great velocity conflicts with the gradual propagation of pulses in an ether. Mangold has the obvious answer to this. The only thing that can be stated with certainty is that we see a star at the moment in which its light comes in contact with the eye. It may nonetheless be possible that the light reaches us at a moment when the star is already some way above the horizon. The fact that this is indeed the case results from the finite velocity of light, something demonstrated by Römer. The last of the two 'rhetorical' objections is that, in conflict with observation, if light is the propagation of a motion in the ether, there would be no weakening of, and no limit to, the illumination. Mangold's answer is that the motion aroused in the ether by a light source is spread over continually expanding spherical surfaces, causing individual rays to become steadily weaker until they are so weak that they can no longer be perceived.[58] The fourth and last objection can be seen as a reworking of one of Euler's arguments against the emission theory, the one concerning the crossing of rays. How can a great many rays cross one another at one point (something we have to accept if we consider how many points we can distinguish in our field of vision) without causing 'confusion'? Mangold devotes extensive, although not always lucid, attention to this problem, unlike Euler in the 'Nova theoria'.[59] Mangold's answer is based upon three considerations. To begin with, he posits that a perceptible moment in time can be divided into many thousands of 'nonperceptible' moments, in other words, that one and the same ether particle can be displaced in many directions in a short period of time. Mangold subsequently remarks that, although a particle cannot move in different directions simultanuously, it can actually receive and pass on various impulses at the same time. Finally he does admit that, in spite of everything, it can happen that some rays cause mutual disturbance. Yet this is not to imply that no orderly light propagation and clear perception are possible, since a ray does not need to act continuously upon the eye. After all, we see a continuous image even when the ray is interrupted, provided the interruption is not too great. This is demonstrated by a circling, glowing coal, which we see as a continuously illuminated ring. Where the last point is concerned, Mangold's argument is analogous to Segner's defense of an emission theory against the objection connected with intersecting rays.

Krafft and Mangold had little influence; after them, the detailed discussion of the long-familiar pros and cons disappeared temporarily from the scene of battle, only reappearing at the end of the 1780s. As we shall see, by then the extensive treatments of the pros and cons partly followed the pattern that had been laid down in outline by Euler. They also partly pursued an entirely new pattern in which the standard arguments played a subordinate role or even disappeared completely.

3. Physical phenomena

a. Rectilinear propagation

The rectilinear propagation of light constituted a touchstone for every pulse or wave theory of light, after Newton in 1672 – and repeatedly thereafter – had advanced the objection that a theory of this kind was unable to explain this fundamental property of light. Before we examine developments after 1746 I will recapitulate the structure of Newton's argument and the answers formulated within the medium tradition in the period 1690–1740.[60]

The crux of Newton's argument lay in the similarities and differences among the propagation of motion in water, air, and the hypothetical light ether. Were the propagation of light to derive from the motion of pulses or waves in a subtle medium, light would spread into the geometric shadow, as occurs in the case of water, air, and every other 'fluid medium'. However, because light does propagate rectilinearly and does not penetrate into the shadow, at any rate not to any great extent, the basic assumption of a medium theory – the belief that light is a wave or pulse within a subtle medium – is incorrect. Light does *not* propagate in the way pulses or waves do, and thus there can be no light ether. In other words, the analogy between sound and light is a *negative* one. The analogy is also an *absolute* one; the propagation of sound is not just somewhat different from that of light; it is completely different.[61]

Huygens provided a partial, that is, *gradual*, and *positive* analogy in contrast to Newton's absolute and negative one. In this way he tried to avoid Newton's conclusion on the impossibility of formulating a pulse theory of light. He did not dispute, in any fundamental way, Newton's empirical evidence about sound waves and rays of light, but he made the difference between the two phenomena gradual rather than absolute. Huygens believed that diffusion occurs in both sound and light propagation. Secondary spherical pulses spread out on all sides of an opening or obstacle, but sound or light is best perceived where the secondary pulses come together in one pulse front. Further, diffusion in the geometric 'shadow' is most perceptible in water waves, is less so in sound waves, and is imperceptible in light waves. The great lacuna in Huygens' response was that he gave no reason for the gradual propagational differences between the three media. Yet it was significant that he drew a gradual rather than an absolute analogy and that he tried to solve the problem on the theoretical level without criticising in any fundamental way Newton's interpretation of the empirical data. Johann II Bernoulli's stand resembled that of Huygens. Although he dismissed Huygens' secondary pulses and the principle underlying them, he also drew a gradual positive analogy between sound and light: Both phenomena are motions within a medium, but the nature of the medium differs. Because of the existence of incompressible colour particles in the light medium, he believed that a light pulse propagates in an exclusively rectilinear way. Thus, the discrepancy between sound and light media is a difference in the elasticity of the two media.

A second kind of response to Newton's objection stemmed from Euler who, unlike Huygens and Bernoulli, used an *absolute* analogy. In order to avoid arriving at the same conclusion as Newton had done, Euler is compelled to adopt the opposite analogy to that of Newton. If the latter conceived of the analogy between light and sound as a negative one, Euler regarded this as a *positive* one, positing that light and sound are both propagated by means of periodic pulses that do not penetrate into the shadow and that propagate in a strictly rectilinear way. Like Huygens and Bernoulli, Euler was contradicting Newton on the theoretical level here, and there was now a clear empirical difference as well: Euler posited, in opposition to Newton, that the propagation of sound is also rectilinear. The weak point in Euler's conception is that a rectilinear propagation of sound appears to conflict with everyday observation, and with the concomitant hypothesis that sound derives from the transmission of pulses or waves in the air, which also spread sideways when encountering an opening or obstacle. It is this circumstance that leads to the remarkable fact that discussion in the second half of the eighteenth century on the rectilinear propagation of *light* often revolves round the question of whether *sound* is propagated rectilinearly.

When we turn our attention to the arguments concerning rectilinear propagation in the years after 1746, we see that in the majority of cases Newton's or Euler's reasoning is adopted with no alterations or additions and sometimes almost completely literally.[62] Thus, the absolute analogy between sound and light, whether viewed positively or negatively, dominated the discussion. Even though Huygens' theory of secondary pulses was rejected within the medium tradition, his idea of a gradual positive analogy between sound and light still survived in places. He was sometimes cited for this reason.[63] However, the small number of people who championed a gradual analogy more often advanced the idea, also appearing in Johann II Bernoulli's work, that there is a difference in the elasticity of light and sound media. We have seen one example of this in the case of Mangold, who thought that air particles and ether particles displayed a lateral spreading that was small in the former case and very small in the latter. Yet Mangold made no mention of Bernoulli's elaboration of this idea (colour particles, and so on). Reduction of the propagational difference between sound and light pulses to a variation in elasticity was probably such an obvious idea that nobody should have felt obliged to mention Bernoulli in this context.

Mangold's remark on the difference in the elasticity of air and the light medium merely shifts the problem to another point and is no more than a suggestion about the direction in which a solution may be sought. The rudimentary character of this 'solution' is clear in Segner's contribution to the discussion, published in the second edition of his *Einleitung in die Natur-Lehre* of 1754. Although he calls Euler's use of the analogy between sound and light profound, unforced, and natural, he perceives a difficulty in explaining, by means of the wave theory, how a vibrating medium can only propagate its motion in a rectilinear fashion: 'Soll also diese Lehre behauptet werden, so muss die Elasticität der dünnen Materie, in welcher das Licht fortgesetzt wird, von einer ganz andern Beschaffenheit seyn, als die Elasticität der Luft, und nur vor sich, nicht aber nach der Seite würken'.[64] Segner draws neither a strictly positive nor a strictly negative analogy here. His solution (or rather, his question) follows the line of Bernoulli's hypothesis and Huygens' gradual (positive) analogy. The difference in elasticity between the two media would explain the uncontroversial empirical differences in propagation.

Unlike Mangold and Segner, Johann Heinrich Lambert disputed precisely Newton's conception of sound propagation. He did this in an article of 1770 on the speaking-trumpet, an instrument that can carry

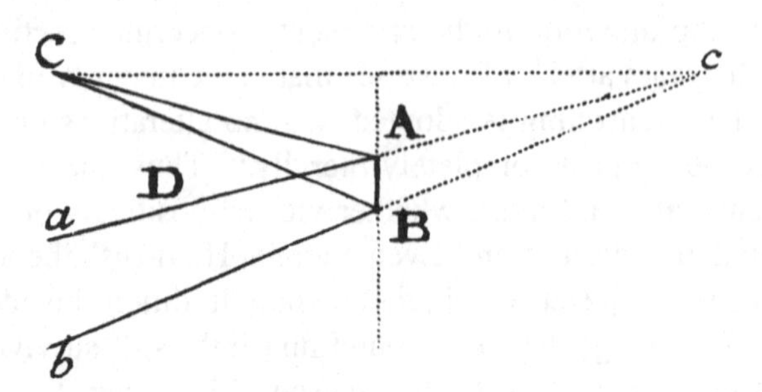

Figure 11. Lambert's echo experiment (Lambert, 'Instrumens', Pl. V, Fig. 1).

the spoken word over a considerable distance, used by seamen especially. Lambert conceived of the speaking-trumpet as a composite acoustic mirror in which 'sound rays' are reflected. On this assumption the familiar theory of mirrors has only to be applied to find the ideal form for a speaking-trumpet. However, the problem is a complex one, since the source cannot be assumed to be a point source; there are many reflections; and there is resonance, as in the case of musical instruments.[65] Before he begins the discussion proper of the speaking-trumpet, Lambert wants to render acceptable the idea that the supposed positive analogy between sound and light is a correct one. It is this introductory passage that is directly concerned with our subject.

Lambert observes that it is virtually impossible to produce a sound that propagates in only one direction. He subsequently questions whether this is also fundamentally impossible. According to Newton's theory of sound, it does seem to be impossible: '*Newton* . . . rapporte même une Expérience pour constater la chose, & pour prouver en même tems, que le son & la lumiere n'ont rien de commun pour ce qui regarde le mécanisme de leur propagation'. He discusses the 'one-hole observation' meant in this quotation, and sides with Euler, who did not regard Newton's reasoning as decisive to the matter. Lambert subsequently says: 'Je n'ai pas fait l'expérience proposée par *Newton*, mais je sais, par d'autres observations analogues, ce que cette expérience peut faire voir; c'est que l'oreille placée dans le cone sonore [sound cone] dont je viens de parler, entend plus clairement & plus fortement, que lorsqu'il est de côté'. One of these 'autres observations' concerns the phenomenon of the echo. Lambert treats of this as an example (see Figure 11).[66]

Lambert's starting point for his discussion of the echo phenomenon is as follows. There is a rockface *AB* of moderate size and at a sufficient distance from the sound source. In cases of this sort he always observed that the echo could not be heard everywhere in the surroundings, only in the places to which reflected sound went in accordance with the rules of reflection. He corroborates this general remark with several specific observations (see Figure 11). Between *a* and *b* he heard the sound that in fact came from *C*, as though it derived from *c*. He also perceived that the distance *ab* was fairly small. Where the obstacle *AB* was curved, *ab* was sometimes even smaller than *AB* and the echo intensified the sound, just as a concave mirror would concentrate rays of light. Evidently it is possible to assume that sound propagates in rectilinear fashion, an idea that conflicts with Newton's conception, Lambert concludes. In Newton's view a particle of air moving in one or another direction must set neighbouring particles in motion; in this way sound would spread in all directions. Yet, Lambert maintains, echo phenomena show that Newton's conception is incorrect. If the particles *D* just outside the 'sound cone' *aABb* were to become involved in the motion, then that motion must be 'extrèmement affoibli', because the echo was only audible when the ear was actually inside the cone.[67]

Lambert is the first after Huygens to use the echo as an argument in the discussion on rectilinear propagation. Like Huygens, he provides virtually no quantitative data about his observations. His perception of the intensification of sound and the diminishing of the 'cone width' *ab* does constitute a step in that direction; at all events Lambert's observations are more precise than the remark by Huygens. A fundamental difference from the latter is that Lambert makes no use of a pulse front construction, and that he therefore deals in a different way with the echo argument. Huygens utilises it to underpin his *gradual* positive analogy, whereas Lambert locates it within the context of an *absolute* positive analogy. The result is that Lambert, like Euler, has to contest Newton's theory of sound. Euler restricts his attention to the passage of sound through an opening, whereas Lambert discusses a complementary situation, reflection from an obstacle.

Nicolas Béguelin, in a lecture given to the Berlin Academy of Science, published in 1774, investigated the possibilities of arriving at a decision experimentally in the discussion between emission and medium traditions.[68] After an enumeration and elimination of the potential candidates for an *experimentum crucis*, two remained. The first concerns the possible refraction of 'grazing' rays of light (those with an angle of incidence of 90 degrees), which will be discussed in the next subsection. The second experiment nominated refers to rectilinear propagation and is dealt with here.

Béguelin regards Euler's solution to the problem of light and sound propagation as 'plus ingénieuse que solide'. He poses a number of questions with a critical intent, concerning Euler's idea that sound pulses are propagated rectilinearly. Why does sound appear to grow louder in a curved tube? Why can one hear sound-producing bodies that cannot be seen? Béguelin does not regard these observations as decisive, since one or another explanation can be found for every difficulty, within the framework of Euler's wave theory. Multiple reflections seem to him especially well suited as the explanation of the lateral motion of sound, which is apparent only in Euler's wave theory. Béguelin discusses a potentially crucial experiment connected with Euler's ideas on a room with sound-proof walls. However, he makes it clear right from the start that for him, too, this is a thought experiment, his principal intention being to provide an elucidation of the discussion on rectilinear propagation.[69]

Béguelin's proposed investigation was only a thought experiment because technical difficulties prevented it from being performed; the material required was one that would absorb sound completely and would therefore give absolutely no perceptible reflection. The ceilings, walls, and floors of two rooms connected by a door opening had to be covered with this – unobtainable – material. Béguelin was contemplating what we now call anechoic rooms.[70]

Let us now suppose that a sound is made in the first room. If sound were propagated rectilinearly (Euler's conception), there would be completely silent places in the second room, because these spots would lie in the acoustic shadow of the connecting door. If, on the other hand, sound propagated in every direction (Newton's opinion[71]), there would be no spot in the second room where nothing could be heard. Béguelin formulated a clear conclusion for the case of the second situation (no silent places): '[I]l semble qu'on sera en droit de conclure que la propagation du son differe essentiellement de celle de la lumiere, & le systême de l'émission paroîtroit en quelque maniere démontré.'[72] It is clear that in such a case Béguelin sees no hope of supporting the medium conception of light by an auxiliary hypothesis.

This claim requires some elucidation. Elsewhere in his article Béguelin avoids hard and fast conclusions. In these places his judgement is balanced and cautious, harmonising with his general philosophical attitude, which can be characterised as undogmatic and conciliatory.[73] One does not expect a firm pronouncement from a man like Béguelin on a possible refutation of the medium tradition. In my opinion, part of the solution to this paradox lies in the 'ontological conception' of analogies that was widespread in the eighteenth century. In this conception there is agreement between two phenomena if these have a common physical explanation. If light is analogous to sound, then there is an ether as the material medium for the propagation of pulses or waves. Moreover, the properties of the ether are approximately the same as those of air, the medium within which sound is propagated. Consequently, if the propagation of light in the ether is rectilinear and that of sound in the air is not, the basis for the analogy disappears. An important empirical difference between light and sound is translated directly into a fundamental difference on the ontological level, and the two phenomena are irrevocably divorced from each other. The ontological concept of analogies does not permit any partial or gradual analogy: Either light is analogous to sound, or the two phenomena are not analogous at all.

In 1772 Priestley published an extended work on the history of optics and its current state. In discussing Euler's theory of light, Priestley makes implicit mention of the argument concerning rectilinear propagation, recalling that to most natural philosophers the arguments against a medium hypothesis appeared to be decisive, 'particularly after the publication of Sir Isaac Newton's *Principia*, in which the impossibility of that hypothesis seemed to have been demonstrated'.[74] A German translation and adaptation of Priestley's work appeared in 1776. The translator and annotator, Georg Simon Klügel, added one or two pages to Priestley's brief description and evaluation of Euler's theory.

Among other matters, Klügel mentions Euler's response to Newton's objection concerning rectilinear propagation. As we have seen, Euler believed that, in a room with sound-proof walls, a noise would only be audible in straight lines behind an opening. Klügel – who incidentally makes no mention of Béguelin's elaboration of Euler's idea, although he was aware of it[75] – claims that he has actually succeeded in performing the experiment that Euler had thought out but regarded as impracticable. Klügel asked someone to read aloud from a book, outside his room, directly opposite the door, which was at a distance of approximately 6 metres. Klügel himself sat in the room, to one side of the door and at a distance of 5 metres from it. The line linking him with the reader passed through a chimney and two thick walls enclosing a spiral stone staircase. With the door open he was just able to understand what the reader was saying, but with the door closed all he could hear was a faint mumble. Similarly, where in the first instance he could barely understand what the reader was saying, in the second instance he could only occasionally hear an almost imperceptible sound. Klügel concludes from this that sound cannot have entered his ears via a straight line, since in that case there would have been no noticeably great difference. The situation was also such that the sound could not reach him through the open door in a simple reflection and that multiple reflections were not feasible 'wegen des in dem Zimmer befindlichen, hiezu sehr untauglichen Geräthes'. The sound must therefore have propagated in a lateral direction from the door. Klügel agrees with Euler that the listener thinks that the sound does not come from the door opening, but in his opinion this occurs because the ear's judgement is much less reliable than that of the eye.[76]

In this context Klügel pronounces his judgement on Huygens' pulse theory. He outlines Huygens' explanation of rectilinear propagation by means of spherical pulses and pulse fronts. His way of doing this is not entirely comprehensible, something that might arise from his own incomprehension of the subject or even from repugnance. 'Es gehöret . . . nicht wenig Mühe dazu', Klügel complains, 'seine Einbildungskraft der Hugenianischen gleichförmig zu machen, mehr als die ganze Sache werth ist'. His judgement of Huygens' explanation is consequently extremely critical: 'Es mag aber wohl mit diesen Nebenwirbeln [secondary pulses] nicht besser beschaffen seyn, als mit den Epicyklen in dem Ptolemaischen Weltsysteme. Solche in eine Hypothese hinein gezwungene Flickbegriffe [literally 'patch-up concepts'] erregen kein gutes Vorurtheil für dasselbe'.[77] The rejection of Huygens' principle was, as we have seen, a general phenomenon in the eighteenth century, with only one or two exceptions at the beginning of the century. Here, as in other cases, this serious candidate for a gradual analogy between sound and light was shown the door.

Klügel is thinking in terms of an absolute analogy. In his case this was a negative analogy, in contrast to those of Lambert and Euler, but in accord with Newton's. Sound propagates sideways, whereas light does not; therefore, light is completely nonanalogous to sound. However, one cannot conclude from Klügel's critique of Huygens and Euler that he subscribed to an emission theory. He observes in his notes and appendices that problems attend both optical traditions, and he declines to elect either one or the other. This neutral position stems from his conviction that the nature of things – in this case, light – is unknowable for humankind. He conveys this point of view, which resonates through many of his comments, at the end of his appendix on Euler, by means of Plato's metaphor of the cave, although he uses this in a different way from Plato: Physics is the theory of the shadows of things, rather than of the things themselves. In a note to the section heading 'Summarische Vorstellung der Lehre vom Lichte' he says: 'Ich nehme an den hier vorgetragenen Hypothesen keinen Theil. Hypothesen sind nur Bemäntelungen unserer Unwissenheit'.[78]

Euler's hypothesis that sound propagates in a strictly rectilinear manner; Mangold's and Segner's suggestions concerning a gradual difference in elasticity between the media of sound and light; Lambert's experiment in support of Euler; Béguelin's thought experiment; Klügel's experiment in disproof of Euler – none of these produced a decision in the debate on the nature of light, not even in the disputed argument concerning rectilinear propagation. This is obvious where Euler's hypothesis is concerned: At first acquaintance it appears too unnatural and requires, at the very least, further elaboration; the suggestions on gradations in elasticity remain too sketchily elaborated, and Béguelin's contribution concerns a thought experiment that serves only to elucidate the question itself. Why, however, were Lambert's or Klügel's experiments not decisive? Lambert's analysis of the echo phenomenon could only be decisive in respect of the controversy on the rectilinear propagation of sound. It functions only as a defensive argument in the discussion on light. Its aim was to neutralise Newton's objection and could therefore at best salvage the candidacy of a medium theory. It was unable to demonstrate the incorrectness of an emission theory. Klügel's experiment had potentially farther-reaching consequences. He came close to the conclusion that Euler's theory could not explain rectilinear propagation and was therefore very implausible. However, he did not state this deduction in any explicit way, perhaps because of his sceptical attitude. We do not know whether others actually drew this conclusion. Those who would have been most likely to react – Euler and Béguelin – expressed no opinion on Klügel's observations, or at any rate not in print.[79]

Euler's hypothesis on the propagation of sound and the subsequent discussion brought no solution to the problem of the way in which the propagation of light ought to be explained within a medium theory. Why is this? One could simply posit that the great 'error' of eighteenth-century theorists in the medium tradition was their failure to recognise that the dimensions of the experimental arrangement were relevant to whether there was a more or less significant deviation from rectilinearity. The situations treated as being analogous differed considerably on this point. In the optical case – a (not too) small hole in a window blind – the wavelength of the light used was very much smaller than the dimensions of the opening. Consequently the effects of diffraction were limited and these, if they were observed at all, could be regarded as minor deviations that did not detract from the fundamentally rectilinear propagation. These deviations required further explanation, which in the emission tradition was usually a short distance force. In contrast, the dimensions of the opening in the acoustic situations usually cited were of the same order of magnitude as the wavelength of sound. In that case there was considerable deviation from the straight path. In Lambert's echo experiments, however, the measurements of the obstacle were probably greater by one or two orders than in the room door experiment; therefore, in Lambert's case spreading into the shadow was noticeably reduced.

There are one or two presuppositions in this response to the question of why the problem of rectilinear propagation was not solved within the eighteenth-century medium tradition, and it is therefore no real answer, at best the beginning of one. Connecting wavelength and rectilinear propagation means, after all, that the implicit assumption is Fresnel's theory of diffraction, in which wave interference and Huygens' principle are linked. As we now know, this theory deriving from the first half of the nineteenth century offers a solution to the problem.[80] The answer given above therefore amounts to an assertion that an adequate theory of diffraction had not yet been devised. There still remains the question of why this theory had not been discovered by Euler or one of the participants in the discussion. In my view there are several interlocking factors in play here. In consecutive order these are the nature of the problem; the absence of, or insufficient presence of several essential theoretical concepts; and the weakness (not only within physical optics but also within physics as a whole) of a research tradition binding mathematics, physical theory, and experiments.

Let us begin by examining the formulation of the problem. A veritable devil's plot seems to have been hatched between Newton and Euler. Both Newton and Euler regarded the rectilinear propagation of light as being the primary phenomenon, making it difficult if not downright impossible to reach Fresnel's solution to the problem, or any other solution in which rectilinear propagation is a derived phenomenon. In the open field, light was propagated rectilinearly, only deviating to a small degree from the straight line if it ran close to objects. The refraction of light is thus secondary. This manner of posing the problem largely determines the direction in which a possible explanation was being sought. The uniqueness of Huygens' theory lies in the fact that the relationship is reversed, the primary factor being that light pulses extend to every side. For him, the rectilinear propagation of light is a derived phenomenon. However, it was precisely Huygens' theory that was categorically rejected both within and outside the medium tradition. There is a second aspect of the problem construction, connected with the factors just discussed, to hinder arrival at a solution of the Fresnel type. Because the idea of rectilinear propagation was taken to be primary, it was tempting if not actually unavoidable to pose the problem in terms of two mutually exclusive options: Either light propagated in rectilinear fashion, or it did not. In other words, there were sharply defined alternatives and no middle ground. This was reflected in, or it was stimulated by, the absolute and ontological use of the analogy between sound and light by both supporters and opponents of a medium theory, as we regularly observed above. There was virtually no partial or gradual analogy and even where it did occur, as in Mangold's and Segner's thought, there still remained the first threshold, the absence of the 'correct' relationship between rectilinearity and deviation from it, not to mention the as yet rudimentary character of the gradual analogy.

The availability of theoretical concepts played a role, too. Two ideas were required for a solution of the Fresnel type, and they were either no longer available or not yet devised. The first is Huygens' principle, which was repudiated early in the eighteenth century. In contrast, the second, the idea of interfering waves, was certainly in the air but was only poorly articulated and was not linked with optical theory.

Finally, the style of investigation is significant. There was no lack of relevant optical experiments in the eighteenth century. Many tests were carried out on the inflection of light,[81] yet nobody arrived at the idea of regarding inflection as the norm and rectilinear propagation as the deviation. The experiments into the propagation of sound were often intended to measure the velocity of sound. Sometimes they were, it is true, related to rectilinear propagation, and they were used in the discussion on light, yet in the cases we have seen here only the strictest of alternatives were granted consideration. Nothing, however, or barely anything, was quantified in either kind of experiment. This is symptomatic of a deeper-lying factor, one I shall only mention here. This is the lack in physical optics, of a research style in which mathematics, physical theory, and experiments are blended into a harmonious whole. I shall return to this point in the epilogue.

b. Grazing rays of light

Although rectilinear propagation was the physical phenomenon most frequently cited in the discussion on light, there were other physical phenomena on the agenda. One of these will be treated as an example here, because it provided fresh evidence in the debate and because it was explicitly designed as a crucial experiment. The phenomenon under examination concerns the question of what happens if a ray of light grazes a glass surface. Is a grazing ray refracted, or does it continue unaffected along its original straight path? This query and its accompanying experiment occurs in Béguelin's article of 1774, mentioned above, published in the memoirs of the Berlin Academy for 1772.

Before Béguelin reports on his experiment concerning grazing rays of light, he discusses the theory behind the phenomenon. What do the various theories of light have to say about the case of grazing rays? According to 'the' emission theory, an attractive force is exerted on a grazing ray of light that causes the light particles to diverge from their straight path and enter the glass. According to this theory, there should therefore be refraction (Figure 12a). (N.B. Béguelin himself provides no figures.) In 'the' medium theory refraction is explained by the different densities of air and glass, with the resulting different velocities in the two media. Grazing rays of light move parallel to the surface of the glass, thus always remaining in the same medium. There is no reason for a change of velocity or direction. 'Le rayon continuera par conséquent son chemin en droite ligne, il rasera la surface du milieu dense sans la pénétrer en aucun sens'[82] (see Figure 12b). The predictions of the two theories clearly differ, and it is therefore a simple matter to carry out an experiment to show which of the two is the correct one; at least, that is the impression Béguelin conveys at this point of his discussion.

Before giving the results of his experiment, Béguelin comments on the theoretical significance of the potential outcome. If refraction oc-

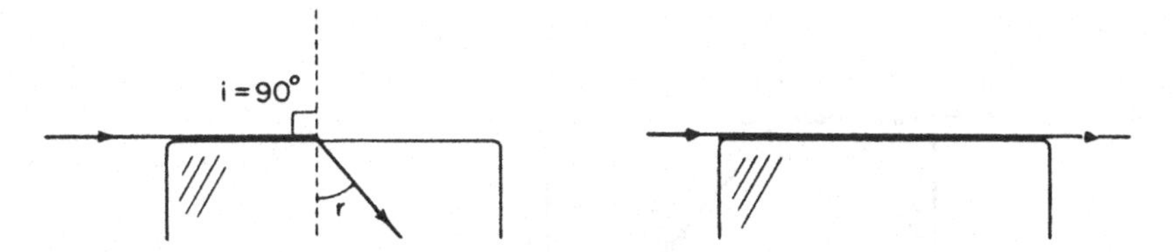

Figure 12. Béguelin's experiment on grazing rays of light. (a) Prediction according to emission theory: refraction. (b) Prediction according to medium theory: no refraction.

curred, then two propositions from 'the' emission theory would be confirmed: first, that light consists of emitted particles and, second, that refraction is the result of an attractive force. If no refraction could be demonstrated in the experiment, the conclusion would be less clear, since it would be shown only that refraction is not caused by an attractive force. However, the emission theory in its entirety would not consequently be rejected. In particular, light could still consist of emitted particles.[83]

Béguelin arranged his experiment in a simple way. He had an elongated container made, approximately 40 centimetres long, closed at the ends and open on top. A glass cube of 5 to 10 centimetres (a few inches) in each direction could be placed in the container. Solar rays entered the room through a small horizontal slit. The closed end on one side of the container prevented light from entering the side of the glass cube. A horizontal line was drawn on the other end of the container. This 'gauge line' would be illuminated if there was no refraction when the rays of light entered in grazing incidence; that is, the gauge line would be illuminated in the case of rectilinear propagation. The remainder of the container was blackened, and paper was laid under the cube to enable the investigator to see if light left the cube there. In performing the experiment Béguelin placed the container with its cube on a table, raising one end of the container until the rays of light skimmed the upper surface of the cube.[84]

In discussing the possible results of his experiment, Béguelin makes a mistake in calculating the anticipated path of the light ray if it is refracted on the upper surface of the cube (see Figure 13). He thinks that a grazing ray of light would follow the path *ABCD*, whereas a little geometry and the laws of refraction and reflection lead to the path *ABCEF*.[85] The conclusions Béguelin draws from his experiment do

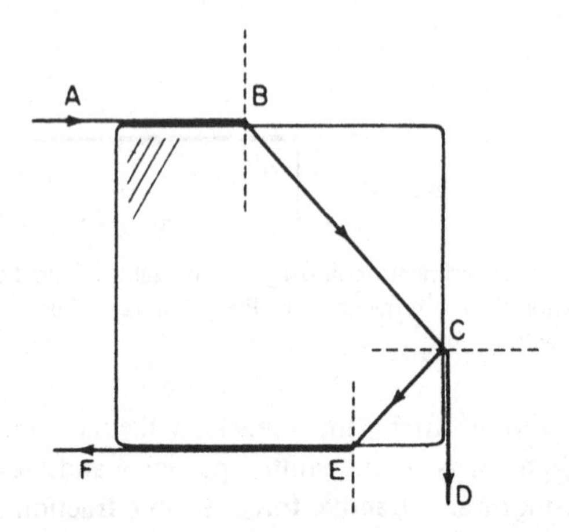

Figure 13. The path taken by a grazing ray of light (if refraction were to occur): *ABCD* is the path according to Béguelin, whereas *ABCEF* is the correct path.

not have to be changed if this correction is made. Nevertheless, the credibility of his results was adversely affected, partly because an influential critic focussed attention on his error (see below).

Béguelin performed the experiment with grazing rays of light several times, always obtaining the same results.

If the angle of incidence i (see Figure 12a for the definition of i) was smaller than 90°, then refraction occurred. This was deduced from the perception of light on the paper lying beneath the bottom surface of the cube.[86] When $i = 90°$ (grazing rays), there was no refraction. All the light at the bottom disappeared, and the gauge line on the end of the container was illuminated.[87]

The second result led Béguelin to conclude that refraction is not caused by an attractive force. He even made this conclusion more general: According to Béguelin, the attraction and repulsion of light, as well as the fits of easy transmission and reflection, were at best facts perceived by Newton, but still lacking theoretical explanation.[88]

It was no easy task to explain, once again, such diverse phenomena as reflection, refraction and Newton's rings, and it is thus easy to understand why the supporters of the emission tradition did not rise to Béguelin's challenge. There were other reasons, as well, and in order to discover what these were, I shall first deal with Béguelin's discussion of potential objections and then review some of the comments on Béguelin's article made by others.

The first possible objection to which Béguelin turns his attention concerns the observed facts. A defender of the attractive force may think that the ray of light *CD* in Figure 13 is drawn back into the glass and refracted. It could also occur twice, the net result being that light, after being refracted four times over, would continue in a straight line. If this were true, the gauge line would be illuminated by rays of light entering at a grazing angle, while refraction would have nonetheless occurred. (Béguelin's error concerning the light path resurfaces in this comment, but if we correct this the argument remains essentially the same.) In answer, Béguelin makes several predictable comments – for example, that light would have to be visible on the paper at the bottom of the cube if it were to leave the cube there, but such had not been observed. The second potential objection concerns a more theoretical point. One might defend the attractive force by positing that this force is only active at an infinitely small distance from the surface of the glass. Béguelin comments that the expression 'infinitely small' cannot be taken literally. One must either suppose that the glass attracts no rays of light at all or admit that a certain number of rays are attracted and refracted. In the latter case the rays, however few, would have to be visible. These two objections did not affect Béguelin's conclusion that there was no attractive force. To avoid misunderstandings, he repeats that his experiment does not demonstrate the invalidity of 'le systême de l'émission': '[L]a principale question, celle qui roule sur la maniere dont la lumiere est propagée, reste encore indécise'.[89]

To sum up, we can say that Béguelin's insignificant error concerning the path of light in the glass cube undermined the credibility of his conclusion. Furthermore, he gave little emphasis to the far-reaching consequences his experiment had for the debate on the nature of light.

There were many responses to Béguelin's experiment, the first and most influential being anonymous. This constituted part of a review in the *Allgemeine deutsche Bibliothek* of the memoirs of the Berlin Academy for 1772, in which Béguelin's article appeared.[90]

The anonymous reviewer provides a summary of Béguelin's article and a detailed examination of the experiment with grazing rays of light. He draws attention to Béguelin's error in the matter of the path taken by the rays of light in the glass cube. The critic subsequently observes that although application of the law of refraction suggests that one can indicate a point on the upper surface of the glass cube at which the perpendicular can be drawn (see Figure 13), if the angle of incidence is 90°, the perpendicular can be drawn at every point of the upper surface. However, '[N]ach dem Satze des zureichenden Grundes, den Hr. B. ohnedem gern anwendet, sollte man muthmassen, der Strahl werde sich in keinem dieser Puncte brechen, weil er sich nicht in allen zugleich brechen kan'. According to the reviewer this shows that the manner in which the law of refraction has to be applied when $i = 90°$ 'noch etwas dunkels hat'.[91] This reservation creates a rather remarkable and abstract mathematical impression; it does not take sufficient account of the physical circumstances of the experiment. Finally, the reviewer posits that Béguelin's pencil of solar rays cannot be regarded as one ray coming from one point. In reality the rays in the pencil are not parallel with one another. The critic concludes by saying that his comments are not intended to refute Béguelin's results and conclusions, but rather to ensure that the experiment will in future be repeated in a more satisfactory way.

The experiment was not repeated, but the reviewer's criticism was. In Germany the review had a decisive influence on the reception accorded to Béguelin's experiment. This emerges clearly from Klügel's reaction to the experiment in his edition of Priestley's history of optics. In a brief summary of Béguelin's article, Klügel gives his opinion on the experiment with grazing rays: '[Béguelin] erzählet . . . einen von ihm angestellten Versuch, der billig schon längst hätte müssen gemacht werden, dadurch die anziehende Kraft, welche in dem Newtonianischen System die Brechung erklären muss, umgestossen wird'. Here Klügel concurs with Béguelin's conclusion, but he changed his mind when he read the anonymous critique, adding a note in an appendix on his own positive comment on Béguelin's experiment. Drawing attention to the reviewer's 'gegründete Erinnerungen', he says that the manner of applying the law of refraction 'freylich noch etwas dunkles [hat]'. This is the formulation used by the reviewer. However, Klügel does not appear to be basing his judgement primarily on the abstract mathematical argument about the place of the perpendicular, but on a physical restriction to the law of refraction. He concentrates on the phenomenon of partial reflection, which is ignored in the usual deductions of the law of refraction.[92] Three other statements in the German literature have the same tenor as Klügel's comments in the appendix. The thesis that glass possesses no attractive force was mentioned in the same breath as the 'sehr gegründete Erinnerungen', the 'gegründete Einwendungen', and the 'wichtige Erinnerungen' of the anonymous reviewer.[93]

Béguelin's article also became known abroad, even being printed in its entirety in the *Observations sur la physique*.[94] The article was also cited in a discussion on the nature of light in the same journal. This discussion was between Jean Senebier, a defender of an emission theory, and Baron Marivetz, who supported a medium theory. It is clear that only Senebier had read the article, but he used exclusively those of Béguelin's comments that supported an emission theory, thereby creating the impression that Béguelin was on his side. In contrast, his opponent, Baron Marivetz, made no use of the ammunition that Béguelin offered him.[95]

Why was so little value placed on Béguelin's conclusion that there is no refracting force? There are at least three reasons that could be given for this. First, Béguelin did not present his experiment and the thesis based upon it in as convincing a manner as he might have done. The second reason is connected with this. The critique of the reviewer in the *Allgemeine deutsche Bibliothek* probably owed much of its influence to the fact that it caught Béguelin out in an elementary error. Finally, it is telling that a supporter of an emission theory could ignore with impunity the experiment on grazing rays. Senebier cited only those of Béguelin's comments that were favourable to an emission theory. The article could only too easily be used to support either of the two optical traditions.

Béguelin's critique of 'Newtonian' emission theories had little influence. As far as we know his article was not used by eighteenth-century opponents of the emission tradition or, later on, by such people as Young and Fresnel. Jean Baptiste Biot, a major representative of the emission tradition in the first half of the nineteenth century, said in passing in his *Traité de physique* (1816) that a ray of light is refracted most at an angle of incidence of 90°, ignoring the absence of direct evidence for this claim as pinpointed by Béguelin's work.[96]

4. Colour and phosphorescence

In the eighteenth century a phosphorus was a material or object that produced light in the dark without glowing or undergoing a noticeable change. Fireflies, polished diamonds, and rotting wood were well-known producers of 'cold light'. Besides these natural phosphorescent objects and materials, several artificial phosphori were known, one of the most important of the latter being the Bologna stone, discovered at the beginning of the seventeenth century. This phosphorus was made of heavy spar found in the neighbourhood of Bologna (barium sulphate, rich in sulphur), a material that was not naturally phosphorescent, but only after 'calcination' (being heated over glowing coals). The stone also had first to be exposed to sunlight before it gave out light in the darkness. Nowadays we speak of a material with photoluminescence, but in the eighteenth century the term solar phosphorus was used.[97] Solar phosphori entered the German discussion on the nature of light from 1775 onwards.

For what follows later in our narrative, the much stronger phosphorus prepared by John Canton in 1768 is also of importance. Canton's phosphorus was prepared by calcinating oyster shells and grinding them to a powder, which was then heated with added sulphur. A variant of the Canton phosphorus is that used in 1775 by Benjamin Wilson; it was also made from oyster shells. Wilson's investigation particularly concerned the relationship between the colour of the incident light and the light given off by the material.[98] It was Wilson's results in particular, published in book form, that many (including Euler) used as ammunition in the debate on the nature of light. However, this part of the optical discussion cannot be viewed apart from the experimental results and theoretical conceptions before 1775. A short account of all this is therefore in order.

The first experiments on colour and phosphorescence date from 1652 and were performed by N. Zucchi, who believed that phosphorescent light always remained the same regardless of the colour of the incident light. This result was confirmed by an investigation carried out by F. M. Zanotti, the results of which were published in 1748. Zanotti always saw a weak, whitish light, whether blue or red incident light was used. He concluded that the Bologna stone radiates its own inherent light, which is only 'stimulated' into activity by the incoming light. In contrast, Giambattista Beccaria reported in 1771 that in his experiments with the Canton phosphorus the colour of the phosphorescent light was always the same as that of the incident light.[99]

There was no less ambiguity at the theoretical level, with various explanations circulating on the phosphorescence phenomenon. Within the emission tradition the two main conceptions were the 'sponge' and 'combustion' hypotheses. According to the first, the phosphorus soaks up light matter like a sponge, holds it for a while, and then gradually releases the same light particles again. Names for a phosphorus such as light absorber or light magnet are based on this conception. In the second hypothesis the incident light gives rise to slow combustion in the phosphorus, without any noticeable heating effect. This burning causes light particles to be released, with a colour specific to the type of phosphorus. In the combustion hypothesis a chemical process underlies phosphorescence.[100] The sponge hypothesis fits Beccaria's experiments neatly because both incident and outgoing light have the same colour in it. Zucchi's and Zanotti's older results – an inherent light in the phosphorus regardless of the colour of incident light – are difficult if not impossible to fit into the sponge hypothesis, for in this hypothesis the incident and outgoing light particles are the same, and therefore the colour of the emitted light is identical with the colour of the incident light. On the contrary, the combustion hypothesis seems to be confirmed by these experimental results, since the colour of the light emitted may very well be constant, regardless of the incident light particles that function as 'fuel' for the combustion. The theoretical situation is less complex in the medium tradition, where we have only one hypothesis to deal with: Euler's. As we have seen, Euler believed that particles of a phosphorus exposed to sunlight acquire a resonating motion lasting some time, rendering the phosphorus visible in the dark.[101] Euler's conception implies that the outgoing light is the same colour as the incident light, thus harmonising well with Beccaria's results, but not with those of Zucchi and Zanotti.

Yet another factor played a role in the background of the discussion after 1775, that is, the theoretical link between the colours given out by a phosphorus and the 'ordinary' colours of opaque bodies. It is important to observe that, at this point, the dividing line does not run according to the two-way division of emission and medium traditions, but rather cuts straight through it. There is the same sort of theoretical link between the two kinds of bodies in both the sponge hypothesis within the emission tradition and Euler's medium conception. According to the sponge hypothesis, ordinary colours are created by selective reflection of light particles: An object is red because it reflects only the red light particles, absorbing the remaining particles. The phosphorescence colours are explained by assuming a delayed reflection: The particles do penetrate the material, but they are slowly emitted again. A similar theoretical linkage is valid for Euler's conception. The explanation of the colours of an object is in his view based upon the vibration induced in resonating particles. This vibration normally ceases immediately when light is no longer incident on the object. The unusual feature of phosphori is that the vibration lasts some while longer. Both the sponge hypothesis and Euler's explanation use the idea of a delay or extension of the normal process. (It is also obvious in both hypotheses that the phosphorescent light has the same colour as the light absorbed.) Seen from this point of view, the combustion hypothesis is the odd man out, since the behaviour of normal materials and that of phosphori are explained in a fundamentally different way in this hypothesis.

This may be the reason why the combustion hypothesis is entirely ignored as an explanation by some within the emission tradition. For example, in 1772 Priestley claims that Beccaria's experiments demonstrate 'that the phosphorus emitts the very same light that it receives, and no other; and *consequently*, that light consists of real particles of matter, capable of being thus imbibed, retained, and emitted'. Priestley is therefore convinced 'that Zanottus, if he be now living, will retract his conclusion'.[102] In Priestley's view the experiments on colour and phosporescence constitute arguments in support of the emission hypothesis in the discussion on the nature of light. However, he draws the dividing line in a different place from that which one would expect, given what has been written above. Beccaria's experiments could in principle be explained by both the emission and the medium traditions, whereas Zanotti's conclusion *in 1772* could only be fitted into the first tradition. By regarding Beccaria's conclusion as confirmation of the emission tradition, Priestley shows how little he knows about Euler's explanation of phosphorescence – which does match up well with Beccaria's results – or how he fails to take it seriously. Moreover, Priestley apparently does not accept the combustion hypothesis, thereby rendering the emission tradition more vulnerable than is necessary. Shortly after 1772 Beccaria's conclusion was contradicted by Wilson's research, mentioned above. Because of this, the sponge hypothesis and (partly because of Priestley) the emission tradition in general, came under pressure. As we shall see, Euler – who, unlike Priestley, had not yet adopted a definite position in the debate on colour and phosphorescence – did not miss the opportunity of adapting his conception of phosphorescence in a way that enabled him to explain both Beccaria's and Zanotti's conclusions.

What were the experimental results on colour and phosphorescence that Wilson published in his book of 1775? He selected the Canton phosphorus as his point of departure; this always glowed with the same yellowish white, irrespective of the colour of the incident light. By accident he discovered a method utilising oyster shells to manufacture phosphori glowing with the different colours of the solar spectrum. Using both kinds of phosphori, he repeated Beccaria's tests of 1771 but achieved an entirely different result. The colour of the phosphorescent light did not depend upon the colour of the incident light. Referring to Priestley's discussion in 1772, Wilson states: 'I am very happy in having this opportunity of mentioning that the learned Zanotti seems to be much nearer the truth than Dr. Priestley apprehends'. In Wilson's view, phosphorescent light was not the result of the soaking up and subsequent reemission of light matter, but of a '*partial* and *temporary* transmutation between the inflammable parts and the *matter* of light'.[103] In other words, phosphorescence is a combustion, obviously producing barely any heat, that involves a chemical transformation. Wilson is thus speaking out against the sponge hypothesis and in favour of the combustion hypothesis, but he remains within the emission tradition.[104]

Wilson introduced a second element into the discussion in addition to the arguments for the inherent nature of phosphorescent light and against the sponge hypothesis. He observed that in the case of a phosphorus that glowed *red*, the light from the other side of the spectrum – *violet* – caused the phosphorus to glow the most brightly.[105] This result raised a problem both for the sponge hypothesis and for Euler's initial explanation of phosphorescence. After all, phosphorescence was in both cases placed in the same category as the ordinary light of transparent objects, and the expectation was thus that the most effective incident light would be that of the same colour as the light emitted. The combustion hypothesis did not present this difficulty and also seemed to be the most likely way of explaining Wilson's second result.

To attract the attention of the scientific world, Wilson not only wrote a book; he also sent letters to the Berlin and Petersburg academies. This was a successful method since, as we shall see below, he elicited a response from Euler and obtained a great deal of attention in Berlin, with at least three academicians scrutinizing his letter with great care in 1776.

The chemist Andreas Sigismund Marggraf was unwilling to state any opinion on Wilson's results, since he had not had the opportunity of repeating Wilson's experiments for himself.[106] The most detailed response of the three came from the more physics-oriented Béguelin.[107] In his opinion, Wilson's results appear to conflict with familiar observations and conceptions, and in three ways. First, Béguelin states, Newton had assumed that a red body revealed a stronger colour when lit with red than it did under violet illumination, whereas Wilson perceived the exact opposite. Furthermore, since the sequence of colours obtained in refraction of light through a prism appeared to prove that the red rays were more 'powerful' than violet ones, red rays would be expected to cause a red-glowing phosphorus to glow more strongly than would violet rays; however, Wilson observed the opposite phenomenon. Finally, if, as was generally held to be proven (again, in Béguelin's view), the colours of the spectrum were unchanging, a red-glowing phosphorus illuminated with violet light would produce either violet light or no glow at all; Wilson, however, observed that it was precisely the red part of the phosphorescence spectrum that was the strongest in a case of this kind.[108]

Béguelin reported that he had initially placed Wilson's experiments in the category of anomalies – in his own words, 'la nombreuse classe des faits qui sont vrais, mais que je ne puis expliquer'. After some thought however, he maintained that the first two problems could be solved by reducing the differences between Newton and Wilson into differences between a phosphorus and a 'reflecting body' (meaning an opaque body). The latter reflects a portion of the light at the moment of incidence, whereas the former soaks up the rays, retains them for a while in its pores, and then allows them to escape. Béguelin finds it plausible that the red rays penetrate so far into the phosphorus that most of them are unable to free themselves from it, whereas violet rays penetrate less far and are therefore able to emerge again from the phosphorus with greater ease and in larger numbers.[109] Béguelin's idea amounts in fact to an elaboration of the sponge hypothesis. In his view, this hypothesis solved the first two problems but not the last one. After all, the greater *intensity* of phosphorescent light under violet illumination could be explained by the sponge hypothesis, but the difference in *colour* between the incident and outgoing light – a difficulty which Béguelin raised separately – still needed to be explained.

Béguelin believed that a more radical step was needed to dispose of the third problem. In his view Wilson's experiments constituted a reason for doubt concerning the immutability of the colours in prismatic refraction, an essential feature of Newton's theory of colours. The reason for this far-reaching statement was that Béguelin, along with Mariotte and other critics of Newton, was convinced that the colours actually were variable.[110] It is clear from Béguelin's closing sentence that his repudiation of Newton's theory of colours did not mean that he also rejected the emission tradition: 'Je finis par observer que toutes ces Expériences intéressantes de Mr. *Wilson* paroissent plus favorables au systeme de l'émanation de la lumiere, qu'à celui de l'ondulation.'[111] Béguelin had not given any arguments for this statement in his discussion up to that point, and we are thus ignorant of his reasoning. Perhaps he regarded Euler's explanation of phosphorescence as patently incorrect, but it may also be that he was not familiar with it. As we have seen, Béguelin did not take sides in his survey article of 1774 on the nature of light. Nevertheless on that occasion he enumerated several problems attending each of the optical traditions. It would therefore be going too far to interpret the final remark in his report on phosphorescence as a choice in favour of the emission tradition.

In his letter to the Berlin Academy Wilson said nothing about his explanation of the experiments by means of a combustion hypothesis. When the second edition of Wilson's book (which did include this explanation) appeared in 1776, Béguelin added to his first response. Béguelin believed that Wilson conceived phosphorescence as a combustion process in order to maintain the immutability of the colours: The sun's rays ignite a weak flame within the phosphorus, a flame possessing the inherent colour of the phosphorus rather than that of the incident light. In opposition to this, Béguelin says that this does not explain why the weakest rays – the violet – produce the strongest (most intense) flame.[112] Thus, Béguelin is unwilling to embrace the combustion hypothesis without at the same time ascertaining the way his sponge hypothesis, along with the assumption that colours are mutable, explains this phenomenon.

There is a very concise reaction from the third and last Berlin commentator, Lambert. After reading Marggraf's and Béguelin's reactions, he adds: 'Les expériences de Mr. *Wilson* présentent des phénomenes qui tôt ou tard pourront contribuer, sinon à établir la vraie théorie de la lumiere, au moins à renverser de fausses hypotheses.'[113] We may gather from this that Lambert considers the experiments important for the theoretical discussion on light; he leaves open the question of the conclusion that ought to be drawn from them.

Wilson's experiments also evoked a response from the Petersburg Academy. Not only Euler reacted, but also his colleague Wolfgang Ludwig Krafft, who carried out the experiments with the Canton phosphorus, always finding the same white phosphorescent light,[114] confirming Wilson's first results. Euler's contribution is a theoretical one. Enthusiastic as he is, he expresses his views on Wilson's experiments with prismatic phosphori in a somewhat arrogant manner:

> . . . la seule existence de tels corps phosphoriques pourroit suffire à renverser de fond en comble la Théorie Newto-

> nienne sur la lumiere et les couleurs. . . . ces nouvelles expériences . . . sont parfaitement bien d'accord avec ma Théorie de la lumiere et des couleurs, et . . . elles pourroient même servir à la porter à un plus haut degré de certitude, si elle n'étoit pas encore suffisamment constatée.[115]

Euler thus makes it appear as though Wilson's results constitute an unambiguous argument in favour of his wave theory and against the emission tradition.[116] To achieve this, he has to violate the facts on two points. In the first place, he interprets a difficulty concerning the sponge hypothesis as an insuperable obstacle for the entire emission tradition. If we bring in Priestley's polarizing point of view, the same that resonates in Béguelin's response, Euler is right as a polemicist, for Priestley equates the sponge hypothesis and the emission tradition. It is less easy to excuse Euler for a second distortion, since he gives the impression that, even if his wave theory did not actually predict Wilson's results, it can certainly explain them without difficulty. However, he forgets to mention the fact that a marked alteration in his explanation of phosphorescence is required for this, and that his new explanation still leaves many questions open. Just as Priestley had done earlier on, Euler tries to force an either/or choice and to prove that *he* is the one who is right. Yet in each case this is done on the basis of experiments and theoretical explanations that do not permit of such an unambiguous judgement.

How does Euler deal with Wilson's experiments? He begins with an account of the most important result: the unexpectedly marked effect of incident violet light on a red-glowing phosphorus. In his view this presents two problems for the emission tradition. The first is the question of where the red rays come from, since the violet rays are immutable. This is the same vexed question that induced Béguelin to abandon the immutability of colours. The second problem for the emission tradition raised by Euler concerns the phenomenon of phosphorescence in general, rather than Wilson's experiments in particular. Euler recalls that the emission tradition explains the colours of ordinary bodies by the idea of selective reflection. If there is no incident light, there can be no reflection, Euler says, thereby suggesting that the emission tradition cannot explain phosphorescent light in general. He thinks that he can use these two laconic arguments to bring in the crushing verdict cited earlier.[117] This is rather insouciant of him, since in his second argument he acts as though the sponge and combustion hypotheses do not exist.

Euler subsequently provides the adapted version of his own explanation of phosphorescence. He retains his former conception for cases in which both the incident and outgoing light are red: The particles of the phosphorus vibrate because the ether vibrates with the same frequency as their natural one, and they continue vibrating for as long as they are still glowing in the dark. Euler advances a new idea for incident violet light. In this case the phosphorus particles can no longer be induced to vibrate because of the difference in frequencies. Instead, the violet rays induce a 'state of tension' (*état de tension*) in the 'red' resonating particles. As soon as the particles cease to be illuminated, they free themselves from this state and begin their natural vibratory motion. In view of the considerable degree of tension, Euler thinks, the strength of the phosphorescent light is greater than if the body were to be exposed to red rays.[118] This adapted explanation of Euler's leaves many questions open: For example, what are we to understand by a 'state of tension', and why is this tension especially great under violet light?[119] This is not to say that Euler's elucidation compares unfavourably in this respect with the explanations given by its rivals within the emission tradition. The sponge and combustion hypotheses are also vague and ad hoc, lacking any great explanatory power.

Up to now we have not met anybody propounding a combustion hypothesis in reaction to Wilson's experiments. We find one example of this in the natural philosopher Jean André Deluc, who was influential in Germany. His defense of the combustion hypothesis is accompanied by a fierce attack on Euler. Deluc says about Wilson's discovery that a red-glowing phosphorus is activated least by prior radiation with red light, in contrast to prior radiation by differently coloured light, which causes this phosphor to glow more intensely:

> M. Euler triompha un moment de cette découverte; prétendant qu'elle *renversoit* la Théorie de Newton sur les *Couleurs*, & établissoit la sienne sur des *Bases inébranlables*. . . . C'étoit prendre mal son tems pour triompher.[120]

In Deluc's opinion Newton's theory had not been overturned, especially since Newton had expressed an opinion only about 'free' light. Where phosphori were concerned he would certainly have said, with Wilson, that these give out a light that differs from the light coming in. Selective reflection is therefore not the cause of the colours found in phosphorescence. Deluc believes that a certain 'décomposition *phosphorique*' is responsible for the emission of particular kinds of light particles, according to the circumstances obtaining in the illumination and the preceding calcination. It is striking that Deluc does not directly comment on the nature of the colour dependence; rather, he leaves the impression that his combustion hypothesis can explain this. On the other hand, he does quite explicitly bring colour dependence into the arena to do battle with Euler's theory. Deluc cannot accept Euler's explanation employing a state of tension created in a red-glowing phosphorus by illumination with violet light, because in his view Euler's theory predicts that the most intense vibration occurs when the incident illumination is red. Moreover, Deluc says, Euler devoted no attention to the explanation of the white colour of the phosphorus in daylight, since it produces a red light only in the dark. (This is putting the matter in a friendly way, since Euler appears to be assuming, contrary to the facts, that a red-glowing phosphorus is also red in daylight.[121]) Deluc's judgement on Euler's wave theory is just as categorical as Euler's on the emission theories: '[L]'Hypothèse de M. Euler reste alors sans ressource'.[122]

With this, I have outlined the main positions. The theoretical divisions within the emission camp and Euler's rapid ad hoc adaptation of his theory, which allowed both optical traditions to explain Wilson's experiments in their own ways, make it plausible that these tests were not used as decisive arguments in the discussion on the nature of light. It can be added that Wilson's results did not go unchallenged. It was reported that J. N. S. Allemand at Leyden had confirmed Beccaria's conclusion (that incident and outgoing light are of the same colour) in an experiment with the Bologna stone.[123] It was thus possible to retain belief in Beccaria's results even after Wilson's activities. The uncertainty about the experimental results, but particularly the great flexibility with which supporters of either the emission or the medium traditions were able to interpret Wilson's experiments as confirming the validity of their own theories while constituting a problem for their rivals, rendered Wilson's experiments unsuited for a breakthrough in the discussion. This was clear to an independent observer like Gehler. In his authoritative encyclopaedia of physics he left undecided whether Beccaria's or Wilson's experiments produce the right results, although he appears to favour the latter to some degree. However, he thinks that, even if there were greater certainty about the experimental results, they could still not lead to a judgement on the nature of light:

> Man kan aber hieraus [i.e., the experimental results] nicht zwischen Newtons und Eulers Systemen vom Lichte entscheiden, weil sich am Ende beyde Fälle nach einem System eben sowohl, als nach dem andern, erklären lassen.[124]

Gehler failed to include in his discussion the marked effect of violet light on red phosphorescence. But Béguelin's vague hypothesis, as well as Deluc's silence, on the one hand, and Euler's equally unconvincing ideas, on the other, lead us to the same conclusion: The experiments on colour and phosphorescence were unable to end the discussion on the nature of light.

5. The chemical effects of light

a. Changes in the agenda for discussion

Following the section on phosphorescence, Priestley devoted three pages in his voluminous survey on optics (1772) to colour changes caused by light. He reported on experiments into colour alteration in the juice of the murex, the way in which white silver salts become violet, and the fading of flowers. These pages are regarded as the first summary of chemical effects of light.[125] Priestley regards the chemical effects of light as fresh confirmation of the idea that light is 'a real substance' emitted by light-giving objects.[126] If this argument within the treatment of phosphorescence was supported by an explicit hypothesis on the mechanism of the phenomenon (the sponge hypothesis), the reasoning was now implicit and without details. Priestley apparently considers it to be beyond dispute that colour alteration has a material origin, and he feels it to be either unnecessary or impossible to study the mechanism of the chemical effect in greater detail. Both phosphorescence and the chemical effects of light play a minor role in Priestley's presentation, and both kinds of phenomena are never explicitly brought to the fore by him as problems for the medium tradition.

The subordinate role of phosphorescence and the chemical effects of light are easier to see in the German discussion on the nature of light. In the period preceding 1772 neither of the phenomena was advanced as an argument. We have seen the way in which phosphorescence was drawn into the discussion on light. In order to determine how the second topic, the chemical effects of light, came to be on the German agenda for discussion, it is advisable to enquire first of all why it was absent in the year 1772. Two reasons can be given for this absence. First, little was known in 1772 about the chemical effects of light, and the theoretical level was also low. There is, however, a second relevant point, of a disciplinary kind. Even if the chemical effects were to constitute a well-defined area of knowledge, that area would still need to be known, and regarded as relevant to the discussion on light. It is telling that Priestley, also active in the field of chemistry, devoted some attention to the chemical activity of light. However, one might equally well expect nonchemists to pay less attention to a subject of this kind. Since the nature of light was seen as a *physical* problem in 1772, only physical phenomena such as rectilinear propagation and refraction were involved in the discussion. This tacit assumption was demolished fifteen years later; the chemical effects of light were then advanced as a topic for the agenda. In one way this was a minor change; here was one more new argument to add to the already lengthy list. Nevertheless, at the same time it signified a radical alteration: Chemistry was seen as relevant to such fundamental problems as the nature of light. It is the discipline-based character of 'the chemical argument' that distinguishes it from the other arguments in the discussion on light.

What was it that made possible the alteration in the agenda? First of all, a great number of facts about the chemical effects of light were collected and arranged in the years following Priestley's publication.[127] The study of the chemical effects of light profited from the progress being made in pneumatic chemistry (the chemistry of gases), which from the middle of the eighteenth century had constituted an important and rapidly developing area of research within the field of chemistry. With the new knowledge and techniques as a basis, it was possible to study the effects of light in chemical reactions in a more detailed way than before. In the 1770s Priestley carried out a series of tests on the absorption of 'fixed air' (carbon dioxide) and the production of 'dephlogisticated air' (oxygen) by plants. Although he acknowledged the influence of light in this process (which was later to be called photosynthesis), he considered it less significant than did Jan Ingen-Housz, who demonstrated in a book appearing in 1779 that the production of oxygen could only occur under the influence of light. Ingen-Housz also showed that, in addition to the whole plant itself, the plant's fresh parts (such as the leaves) display the phenomenon. He put a leaf in a container with water, using a tube to collect the bubbles of gas forming in the light, and showed that this gas was oxygen. The book in which he published these discoveries, and particularly the experiment with the container of water, caused a sensation, and the experiment became a standard example of a chemical effect of light in the discussion on light. The blackening of silver salts is a second example that was frequently cited. In a book published in 1777 Carl Wilhelm Scheele had provided details of extensive experiments on this topic. In the same work Scheele also published tests on the transformation of calces (metal oxides) into metals in the concentrated light of burning glasses and on the colour changes in nitric acid exposed to light.

In 1782, ten years after Priestley's three pages, there appeared a three-volume book that can be seen as the first authoritative work covering the new field of knowledge. The book, entitled *Mémoires physico-chymiques, sur l'influence de la lumière solaire pour modifier les êtres des trois règnes de la nature, & sur-tout ceux du règne végétal*, was written by Senebier, who not only summarised his predecessors' work, but carried it further with his own experiments. He demonstrated, in a more compelling manner than Ingen-Housz had done, that light rather than heat is essential for the formation of oxygen in green leaves. He also showed, with greater precision than Scheele, that the different spectral colours bring about the blackening of silver salts in very varying degrees. Violet turned out to be the most effective colour.[128]

The research went on steadily after 1782. In 1785 Claude Louis Berthollet made a significant contribution, showing that oxygen bubbles are formed in chlorinated water under the influence of light. Scheele discovered a year later that oxygen is also produced in the photochemical decomposition of nitric acid. This result was confirmed by Berthollet in the same year. In Berthollet's view, light returns elasticity to the 'vital air' bound up in the chemical substance. In other words, light frees the bound oxygen and brings it into a gaseous state.[129] This was not the only hypothesis on the mechanism of the chemical effect of light. Although chemists' opinions differed, all of them agreed that light is a material constituent in a chemical reaction that, like other reagents, has certain 'affinities'; that is, it combines with other chemical substances with varying degrees of ease.[130]

The developments outlined here rapidly became known to the German scientific world. The books and articles written by Priestley, Scheele, Ingen-Housz, Senebier, and Berthollet were quickly translated or summarised in journals.[131] Nonetheless, the summarising and dissemination of knowledge is not in itself sufficient to ensure its use in a context other than its original setting. In this case the new chemical insights had to be seen as also possessing relevance for the discussion on the nature of light, which up to that point had been regarded as a physical discussion and consequently had been conducted almost exclusively in terms of physical arguments.

The activities of people as Priestley and Friedrich Albert Carl Gren in Germany within the field of chemistry helped to ensure that, in the debate on light, attention was paid to photochemical effects. However, there was also a general factor at work. It was at this time – circa 1790 – that there was a considerable change in the status of chemistry in the German universities. Where chemistry had before this time been regarded as a practical field, in some places it now gained recognition as a subject with important things to say about the fundamental structure of the world – about nature's building blocks. On the institutional level this alteration was expressed in the creation of the first professorial chairs in chemistry in the faculty of philosophy; up to that point chemistry had been represented solely within the medical faculty.[132] This disciplinary factor partly determined the force of the chemical argument within the discussion on light.

b. The competition question of 1789

In 1787 the Bavarian Academy of Sciences issued a competition question for 1789, concerning the nature of light: 'Kommt das newtonische oder das eulerische System vom Lichte mit den neuesten Versuchen und Erfahrungen der Physik mehr überein?'[133] Prizes were awarded for two entries, which were published in the Academy's proceedings. The second prize was awarded to Benedict Arbuthnot, an abbot at Regensburg.[134] Arbuthnot, defending an emission theory, appears to have missed the crux of the question. He provides a discussion of the well-known physical arguments for and against the two theoretical traditions, without bringing in new experiments or insights.[135] Arbuthnot sides with Euler, Krafft and Mangold, who – with their standard physical arguments – had been unable to end the discussion on the nature of light.

However, the 'Physik' in the prize question could be interpreted in a wider sense. Of course, 'Physik' was in a wider sense synonymous with 'Naturlehre', or natural science. We can see, from the answer awarded the first prize, that this was the intended concept of 'Physik' and that attention within the natural sciences was directed principally towards the recent developments in chemistry. This prize-winning essay was written by Placidus Heinrich, a Benedictine, later to become professor at Ingolstadt.[136] Heinrich, in his treatise of nearly two hundred printed pages, defends the thesis that the chemical effects of light are crucial to the discussion on the nature of light. In his view, the photochemical effects demonstrate quite unambiguously the correctness of the emission hypothesis of light.

Heinrich's treatise consists of a comprehensive survey of experiments. (He himself performed virtually none of the experiments reported. He was to use many experiments of his own, however, in his later work in the area of luminescence.[137]) Heinrich's outspoken preference for an experimental approach is expressed in his judgement on the theories advanced by Euler and Newton (in the latter case, he is referring to the emission tradition of Newton and the 'Newtonians'). He regards Euler's wave theory as a product of the mathematical mind, whereas he characterises 'Newton's' emission theory as an explanation deduced from facts established empirically.[138]

Heinrich assumes familiarity with the two optical theories and with the objections to them and the responses to those objections. He observes that the arguments for and against the theories advanced up to that date failed to bring the discussion to a conclusion. However, a new era has dawned in the natural sciences, in which

> die Naturlehre durch ihre Verbindung mit der Chemie, durch die Entdeckungen in der Lehre von den verschiedenen Luftarten, von Feuer und Wärme, von der Elektrizität, von der engen Verwandtschaft der drey Reiche der Natur, und ihrem gegenseitigen Einflusse, u.s.w. eine neue Gestalt bekommen hat.

Heinrich believes that he can prove, on the basis of these new observations and experiments, that light

> eine für sich existirende, wirkliche Substanz sey, welche von der Sonne und andern leuchtenden Körpern aussströmmt: dass also Newton im Grunde Recht habe, und das Eulerische System nie werde bestehen können, so lange Beobachtungen und Erscheinungen der Natur etwas gelten.[139]

The argumentation for this thesis diverges to such an extent from what had been advanced in Germany up this point that it is worth reproducing in its entirety:

> Mein Schluss, wodurch ich diesen Satz festsetze, ist folgender: Es lässt sich beweisen, dass das Licht, bloss als Licht betrachtet, eben so gut, und so wesentlich auf andere Körper wirke, nicht bloss nach mechanischen Gesetzen, sondern durch eigenthümliche Kräfte, als sich dieses vom Feuer, von der elektrischen Materie, von den Luftarten, und andern Substanzen beweisen lässt, und
>
> Dass es so entschiedene und sich auszeichnende chemische Verwandtschaften mit gewissen Körpern habe, als es

> von den Säuren, und andern chemischen Auflösungsmitteln gewiss ist.
>
> Wenn man also zugiebt, dass Feuer, Elektrizität, Luft, Auflösungsmittel existiren, so muss man auch zugeben, dass Licht der Körper, und vorzüglich der Sonne, als etwas für sich bestehendes existiret.[140]

In the four chapters of his treatise, Heinrich tries to demonstrate the validity of the thesis with which he begins his reasoning: the existence of nonmechanistic, specific forces of light and, more especially, the existence of special affinities of light for particular substances.

In the first chapter Heinrich discusses the influence exercised by sunlight on the plant kingdom, beginning with widely known observations. Leaves, branches, and whole plants grow towards sunlight, and some flowers open each day in response to light. Plants wither and lose their colour when deprived of light. For these and similar results Heinrich turns mainly to Senebier. He also mentions Senebier's discovery that the violet spectral rays have a greater effect than red rays on the greening of plants.[141] After mentioning Ingen-Housz's experiments on the production of oxygen by leaves in a container of water, Heinrich concludes:

> Es ist ausgemacht, dass Licht auf die Pflanzen wirkt, dass diese Wirkung einem chemischen Prozesse ähnlich ist, dass man sie durch die Verwandtschaften des Lichtes mit den vegetabilischen Substanzen am besten erklären kann: dass also Licht etwas anders ist als Schwingung des Aethers.[142]

In the remainder of his first chapter Heinrich treats of Priestley's experiments on the production of oxygen in 'green matter' (algae). In addition, he provides the reader with several examples of the way in which light affects members of the animal kingdom: the tanning of human skin, the appearance of freckles, the discolouration of silk and wax, and so on.[143] Since these experiments and observations show that heat and light do not always have the same effect, he posits that light matter is not identical with the matter of fire or heat. Further, it is impossible to maintain that light is an elastic medium working mechanically, as Euler would like to make it, because the elasticity and subtlety in a medium of this kind cannot constitute an explanation of the specific chemical effects of light. For Heinrich this proves the first thesis in his demonstration, that light matter exists independently of fire and heat matter and possesses its own kind of force, rather than working in a merely mechanical way.[144]

The second chapter proceeds with an argument, presented in the form of a report on observations and tests on minerals, in support of the second thesis, that light matter has a special affinity for 'pure air' (oxygen) and is able to release oxygen from its bound state and to give it elasticity, in other words, to bring it into the gaseous state.[145]

Heinrich discusses Berthollet's experiments with chlorinated water. Its yellow colour disappears in light, while forming 'dephlogisticated air' (oxygen). Berthollet's experiments with colourless nitric acid had produced similar results: The solution turned yellow, and oxygen was once again produced.[146] Scheele's and Senebier's tests with silver chloride are also discussed. Exposed to light, this material turns first violet, then black. Heinrich also deals with several experiments on the formation of metals from calces in direct sunlight or with burning glasses, carried out by Scheele, Lavoisier, and others.[147] Heinrich believes that all these phenomena can be explained both by the phlogiston theory and by Lavoisier's new chemistry. He declines to choose between them, because he could then be reproached for using one hypothesis to refute another (i.e., Euler's). In his view a choice of this kind is unnecessary, because his arguments against Euler are based only upon phenomena.[148] Heinrich believes that he can conclude, in spite of the varying theoretical explanations of the experiments, that

> das Licht bey sehr vielen Naturerscheinungen als mitwirkende Ursache eintrete, und dass es die vorzügliche Eigenschaft besitze, der reinen Luft, welche in den Körpern gebunden ist, Elastizität zu geben, und sie daraus zu entwicklen.[149]

In a final 'Anwendung der Versuche auf unsre Frage' Heinrich provides one of the clearest formulations of his anti-Euler argument, in which the contrast between chemical and physical or mechanistic explanations is brought to light in a lucid manner:

> Wenn man H. Euler ehedem gesagt hätte, das Licht habe die Eigenschaft, aus den Pflanzen, aus dephlogistisirter Salzsäure, aus Salpetersäure, u. dergl. Luft zu entbinden: es habe die Kraft, verschiedene Metalle ohne Zusatz wieder herzustellen, wie würde H. Euler diese Wirkung, deren Wahrheit sich nicht läugnen lässt, erkläret haben? Den Schwingungen eines überaus feinen Aethers, welche nicht einmahl vermögend sind, die Luft in eine zitternde Bewegung zu setzen – denn der Erfolg davon wäre Schall – hätte er wohl die Kraft nicht zuschreiben können,

> Flüssigkeiten in eine augenscheinliche Gährung zu bringen, die innigst damit verbundene Luft frey zu machen, ja Körper merklich umzuändern. Es würde ihm nichts übrig gewesen seyn, als entweder seinem Aether neue, willkürlich ausgedachte Eigenschaften beyzulegen, oder seine Hypothese fahren zu lassen.[150]

In the third chapter Heinrich states his conception of the relation between light matter and heat matter. His hypothesis is that light matter acts as a kind of solvent for the matter of fire or heat, comparable to the action of an acid on certain substances. According to this idea, sunlight is not of itself hot, but produces indirect heat by means of this 'solvent' action, by rendering the heat matter free and palpable.[151] Since this is only a hypothesis, Heinrich feels that it cannot easily be used against Euler. However, the proposition that heat has a material origin he regards as indisputable, and he sees it as an additional argument against Euler's hypothesis, in which light and heat are conceived of as a motion in the ether and ordinary matter, respectively.[152]

The fourth and last chapter treats of the different kinds of bodies and processes producing light: the sun, electrical phenomena, phosphorescence, combustion. It contains no new features with any importance for Heinrich's central argumentation; it is, rather, an application of the preceding ideas, accompanied by some – clearly incidental – arguments in the discussion on light. For example, the subject of 'electrical light' provides Heinrich with the opportunity for another argument against Euler. He observes that, in Euler's view, the same ether is the cause of both electrical and optical phenomena; however, if, according to Heinrich, Euler's theory of electricity is no longer defended, because it cannot be reconciled with the wealth of new discoveries, there ought to be consequences for the theory of light:

> Man hat längst erkannt, dass der Aether des H. Eulers die physische Ursache der elektrischen Erscheinungen nicht sey, weil sie sich daraus nicht hinreichend erklären lassen; warum soll man nicht eben diesen Schluss in der Theorie von Licht und Farben machen, da eben diese Ursache vorhanden ist?[153]

In the section on phosphorescence Heinrich provides a survey of the discussion that had arisen in response to Wilson's experiments (for this, see Section 4, above). It will occasion no surprise that Heinrich regards these tests as conflicting with Euler's theory and as lending support to an emission theory.[154]

The subjects and arguments of Chapters 3 and 4 are of little value for

Heinrich's main thesis, which is stated and argued in the first two chapters.[155] He claims that, on the basis of observations and experiments, he can positively prove that light is an emission of matter from the sun and other luminous objects and that light matter is a normal constituent of chemical reactions with their accompanying affinities (for oxygen, especially). The objections to Euler's theory are only formulated in general terms. Heinrich does not discuss any detailed wave theoretical explanation. However, this is hardly his fault.[156] The wave theory had barely absorbed the mainly chemical phenomena to which Heinrich was devoting his attention; only phosphorescence had been given some consideration. Moreover, neither Euler nor any of his followers had worked out the theory for the area of photochemical phenomena.[157] Heinrich could thus – without entertaining scruples – be content with varied repetitions of his conviction that an exclusively mechanistic ether could not possibly explain phenomena special to chemistry. His reasoning was backed up by the current state of chemistry and physics, in which nearly everyone had abandoned the attempt to provide mechanistic explanations based upon motion, arrangement, pressure and impulse, sometimes supplemented by forces (such as gravitation) inherent in all matter. The usual explanations relied upon such imponderables as heat matter and electrical fluids: subtle fluids with specific forces or properties.

What are we to think of the originality shown by Heinrich in this treatise? It appears to be minimal if we take account of only the experiments he describes and the conclusions he draws from them. Almost without exception the experiments he enumerates were carried out by others, and Berthollet had already suggested the special affinity of light matter for oxygen. Heinrich's point of view used in the collection and interpretation of observations and experiments – the influence of light on the three kingdoms of the natural world – is not unique and can be found in, for example, Senebier at an earlier date. Senebier had also recognised its relevance for the discussion on the nature of light.[158] Heinrich was therefore original in only one respect, that is, in providing his idea that chemical experiments were not merely important for the discussion on light, but in fact crucial.

6. Chemistry turns the scales

In 1789 Heinrich defended the decisive role played by the chemical argument in the discussion on the nature of light. In the same year Gehler advanced the chemical argument in his *Physikalisches Wörter-*

buch, positing that 'eine nähere Bekanntschaft mit der Chymie jeden für das Emanationssystem geneigter machen müste'. In his view most chemists not only accept the existence of light matter, but use it as an essential component of their best theories. To be sure, Gehler continues, this is still no proof of the actual existence of light matter, since all these theories are only hypothetical, and some of them may prove to be reconcilable with Euler's theory of light. Yet there are phenomena in which light demonstrates affinities for other materials and appears to cause changes in the composition of materials, 'die man schwerlich einem blossen Zittern des Aethers zuschreiben kan'. He provides examples like the production of 'pure air' by plants in sunlight and the discolouration in substances.[159] Here it appears that Gehler was not entirely up-to-date in his knowledge of the scale of photochemical effects. This may be the reason why he denied a decisive role to the chemical argument.

The chemical argument played the same unremarkable part in a physics textbook appearing in 1790. This had originally been written by Wenceslaus Johann Gustav Karsten, professor of mathematics and physics at Halle university, who died in 1787. The 1790 edition of the textbook was prepared for the press by Gren, who was professor of physics and chemistry at the same university. Karsten was a moderate supporter of a medium theory of light; in his view Euler's theory of light had 'viel einnehmendes'. Gren disagreed with this judgement; he maintains in his commentary that Euler's arguments do not refute the emission tradition. Referring to Gehler's article on light, Gren advances as his main objection to Euler's theory the fact that it cannot explain the rectilinear propagation of light. To this he adds: 'Neuere chemische Untersuchungen über das Feuer und seine Entwickelung aus den Körpern, sprechen in der That für das Emanationssystem und für das Daseyn einer Lichtmaterie'. He mentions the same examples as Gehler but adds the blackening of silver salts.[160] Earlier on, in the first edition of Gren's own physics textbook published in 1788, his discussion on the nature of light is less explicit. In this he refers the reader to his lectures for an 'Untersuchung der Gründe und Gegengründe für Eulers und Newtons Meinungen'. Unconnected with this embattled question, he remarks that light matter, like other simple substances, is subject to cohesion with, and affinity for, other kinds of matter.[161] As we have seen, two years later, in 1790, Gren did make the connection between the chemical properties of light and the discussion on the nature of light, even though he was less determined than he might have been.

Up to this point there has been no sign of the direct influence of Heinrich's treatise. Gehler and Gren appear to have obtained their knowledge from foreign sources. Heinrich was only bringing together the discoveries and arguments of others. The only originality he could claim lay in the thesis that the chemical argument was crucial to the discussion on light; up to this point nobody has taken the matter to such lengths. Yet even later, when the chemical argument was being advanced ever more emphatically, Heinrich remained unmentioned, although there seems to be a justifiable suspicion that others used his work.

An influential proponent of the chemical argument was the Göttingen professor Georg Christoph Lichtenberg. His medium was the physics textbook, *Anfangsgründe der Naturlehre*, written by his colleague Johann Christian Polykarp Erxleben, who died in 1777. Lichtenberg saw four editions of this book through the press between 1784 and 1794. In the book Erxleben himself defended Euler's wave theory.[162] The fact that Lichtenberg passed no comments on this point of view in his annotations until 1791[163] does not mean that he had no opinion on the subject. In 1788 or 1789 he wrote a long letter attacking the medium conception. The recipient was Georg Friedrich Werner, who had defended a medium hypothesis in a book published in 1788. Lichtenberg went to great lengths in criticising Werner's volume, using exclusively physical arguments; the photochemical effects did not appear.[164]

Lichtenberg's reasoning was entirely different when he took his stand, in print, against a medium conception, in 1791. In a note he wrote for his third edition of Erxleben's textbook, the chemical effects of light constituted the principle argument for rejecting Euler's wave theory. Here is his witty formulation:

> Das Vibrations-System reicht, vermittelst einiger Hülfsfictionen zwar hin zu erklären wie Helle, Hellheit so entstehen kann wie wir sie bemerken, (und aus diesem Gesichtspunkt ist das Licht bisher fast einzig betrachtet worden) aber nicht, ohne Fictionen mit Fictionen zu häufen und allen Weg der Analogie gänzlich zu verlassen, wie so viele andre Würkungen des Lichts statt finden können.[165]

Although the only example he provides is the unspecified effects of light on metal solutions, he believes that he is able to maintain that, since light has begun to be regarded as matter possessing all kinds of affinities, 'endlich ein Tag in den dunkelsten Gegenden der Physik zu dämmern angefangen hat'.[166] From what he added three years later (directly following this quotation) it is clear that Lichtenberg was certainly not given to understatement, and that he did actually regard an emission theory as the one-eyed man who is king in the land of the blind:

> Hiermit wird aber nicht geläugnet, dass auch diese Vorstellungsart noch ihre Schwierigkeiten habe, und dass wir überhaupt noch weit entfernt sind die Natur des Lichts deutlich zu erkennen, und aus subjektiven Ursachen vielleicht nie ganz erkennen werden.[167]

Despite Lichtenberg's fundamentally sceptical attitude, which prevented him from making a final choice in such matters as the nature of light, it is clear that he, like others, recognised the great force of the chemical argument.

The opposition, in footnotes and additions, to ideas in the main part of a textbook is an intriguing way in which changes in the history of science become visible. The best-known example lies in Clarke's editions of Rohault's physics textbook. In each successive edition Clarke's footnotes contradicted Rohault's 'Cartesianism' in an increasingly detailed and comprehensive way.[168] As we have just seen, this kind of undermining of Euler's optical theory frequently occurred, although the tone of Gren's and Lichtenberg's additions was cautious. The action taken by Friedrich Christian Kries, a mathematics and physics teacher in the Gotha gymnasium, was much more audacious and influential. In 1792 Kries published the first part of a German translation and adaptation of Euler's *Lettres*. Kries apparently regarded short footnotes as an unsuitable way of expressing his criticism of the wave theory of light. He therefore added five completely new letters to Euler's popular exposition of natural philosophy. In these letters he had enough room to explain to his readers (after they had read Euler's objections to the emission tradition) about the worthwhile aspects of the emission tradition and why this tradition was 'generally' accepted.[169]

Kries's beginning is completely right from the psychological point of view. He first deals with several physical arguments against Euler's theory of light, with the aim of weakening confidence in a medium theory. He discusses, for example, the inevitable objection concerning rectilinear propagation and discusses and dismisses Euler's arguments against the emission theories.[170] The stage is then free for a presentation of the basic principles of the emission tradition. Kries restricts his attention to those elements within the emission tradition that can also be found in the rest of natural science. After all, Kries explains, suppose that one theory explained the phenomena of light in the same natural, unforced manner as another theory but also possessed basic principles with a general value; the first theory would then have to be preferred to the second. Ether, the fundamental concept of the medium theories, is something for which we have no further proof. The starting point of the 'Newtonian' emission theory, however, is a universal attractive force (of which gravitation and cohesion are the foremost examples) and the special affinities of light, which are particularly important in chemistry. In Kries's view the universal attractive force is active in refraction, whereas dispersion indicates the action of special affinities.[171]

This is not the only, or even the most important, of Kries's arguments. The last of his supplementary letters is entitled 'Bestätigung der Newtonischen Theorie durch die Chemie' and contains the first definitely stated and fairly comprehensive exposition of the chemical argument after Heinrich. Kries first reminds the reader that light has the power to illuminate and heat, and he does not forget to mention that modern chemists accept the idea of heat matter and reject the notion that heat is a vibratory motion of the particles in a body. Recently, however, says Kries, completely different effects of light have been observed

> die sich unmöglich aus blossen Schwingungen erklären lassen, und die es mithin mehr als wahrscheinlich machen, dass das Licht bey sehr vielen Prozessen der Natur als ein wirklicher Bestandtheil, als etwas körperliches, wirke.

This is followed by an enumeration of the chemical effects of light (which has in the meantime become well known to us), in which three things strike the attention. First, Kries makes no mention of the blackening of silver salts, although his summary is otherwise fairly comprehensive. Second, he examines in detail Wilson's experiments on colours and phosphorescence. Earlier on he had used these to argue against Euler; he can now show that they can be fully explained by the emission tradition. He explains the phenomena by assuming a slow decomposition with its accompanying optical phenomena, in which one colour contained in the incident light will be more effective in producing phosphorescence than another because of its chemical affinity.[172]

Finally, the order in which the chemical examples are presented and the details of the explanation of Wilson's tests indicate that Kries was familiar with Heinrich's treatise. However, Kries did not give a reference to Heinrich's work. Then again, there are virtually no references in Euler's own letters or in Kries's supplementary letters, so a reference to Heinrich's treatise, even if Kries did use it, is not to be expected. This circumstance makes it more difficult for the historian to present a direct link between Kries and Heinrich. The conditions in Germany, the endless discussion between opponents regarded as sharing equal status, and the increased recognition of chemistry as a theoretical discipline, were such that the significance of the photochemical phenomena for the discussion on light, recognised elsewhere, was to be acknowledged here, too, and the photochemical effects were to be decisive in the discussion. With or without a direct link, Heinrich and Kries put into words what, given the circumstances, was very likely to be articulated.

Kries's conclusion clearly formulates the crucial nature of the chemical argument:

> So erhält . . . Newton's Meinung von einer andern Seite her neue Stützen, die der Eulerischen Hypothese gänzlich fehlen. Newton und Euler haben fast nur über die mathematische Möglichkeit gestritten; aber da es hier nicht sowohl darauf ankommt, was mathematisch möglich, sondern was physisch wirklich ist: so gebührt der Chemie ganz vorzüglich die Entscheidung.[173]

Here we find a summary of the group of factors that gives such force to the chemical argument in the German discussion: the recent increase in knowledge in the area of photochemical effects; the fact that these appear to be explicable only by the emission tradition; and finally, the changing status of chemistry, which is claimed to be the highest authority in the fundamental question concerning the nature of light. Kries is pursuing the latter theme when he says:

> Es sind hier auch nach den Bemühungen so vieler der einsichtsvollesten Mathematiker von Seiten der Mathematik nicht leicht neue entscheidende Aufschlüsse zu erwarten; desto mehr aber von Seiten der Chemie, die erst seit einigen Jahren einen ganz neuen Schwung bekommen hat.[174]

After 1792 others also become increasingly convinced of the force of the chemical argument. Gren, for instance, who in 1790 was still formulating his suggestions in a cautious manner, is much more decided in his statements by 1793. In his view the physical objections to Euler's theory of light, coupled with the fact that an emission theory explains the photochemical phenomena in a simple and unforced manner, gives the emission tradition 'ein entscheidendes Übergewicht über die Lehre, dass das Licht nur ein Zustand des Aethers sey'.[175]

The best indication of the extent to which the situation changed within a short period can be found in the fifth volume of Gehler's *Physikalisches Wörterbuch*, a supplement appearing in 1795. The addition to the article on light, published in 1789 (the content has been discussed above) opens with the chemical argument, as Gehler found it in Kries's writings:

> Sehr einleuchtend stellt Herr Kries . . . die Bestätigungen dar, welche die Newtonische Theorie durch die neuern Erfahrungen der Chemiker erhalten hat, und die sich im Wörterbuche [of 1789] . . . nur kurz zusammengezogen finden.

Here Gehler is laying a weak claim to priority, but he does not pursue the matter. He gives a detailed and fairly comprehensive summary of Kries's argument, a discussion that persuades Gehler to alter his judgement on Wilson's experiments with phosphorescence. As we have seen, the experiments with colour and phosphorescence did not seem to favour both of the two optical traditions in Gehler's eyes. He now has an explicit change of heart and underwrites Kries's conclusion that the tests lend unqualified support to an emission theory.[176]

Between Gehler's two articles of 1789 and 1795 there was a complete about-face.[177] The chemical argument gave the emission tradition considerably greater weight than the medium tradition, and at the same time also changed this tradition in two respects. Where content was concerned, the emphasis now lay on light matter as the possessor of qualities and specific properties. Programmatically, change resided in the fact that the direction for further research lay towards chemistry rather than 'mathematics'.

6

Epilogue: Optics as a mirror of eighteenth-century science

In his survey of historical literature on 'experimental natural philosophy' in the eighteenth century, J. L. Heilbron rejects attempts to make the rise of 'Newtonianism' and its ultimate triumph over 'Cartesianism' the guiding historiographical principle. The results reached in the present study on physical optics parallel Heilbron's argument. We have observed that, in the first half of the eighteenth century, the influence of the opposition between 'Newtonianism' and 'Cartesianism' in the discussion on the nature of light, also perceptible, was certainly not a dominant feature. Furthermore, there is confirmation for Heilbron's remark that the disciplinary borders of eighteenth-century experimental natural philosophy, especially optics, were subject to change and that it is precisely these changes that can provide us with a new historiographical guideline.[1]

If we pursue the latter suggestion further, we meet with a general thesis advanced by T. S. Kuhn in 1975.[2] Among all the available alternatives, this thesis is to my eyes the most suited to making the various developments in the eighteenth century comprehensible and to collecting them within one general viewpoint. However, when I attempted to use Kuhn's point of view for the clarification of eighteenth-century optics, I found that I could not use it as it stood. In this epilogue I shall propose a corrective to Kuhn's ideas, one that will be illustrated by material derived from the previous chapters. Before doing so, I shall present Kuhn's thesis itself, and examine the way in which he and others using it have described the development of optics.

1. Kuhn's outline: Mathematical and experimental traditions

The introduction to Kuhn's essay is the most valuable part and at the same time the most obscure. Both the style and the way in which the problem is posed are reminiscent of the Greek pre-Socratics. Just as

Thales, Parmenides, and Democritus presented the problem of change in the world and discussed it in obscure fragments, so Kuhn asks himself how changes in science can be understood.[3] He provides no direct answer to this question, but distinguishes between two historiographical traditions that are relevant to his problem. The first tradition regards science as a collection of separate disciplines – mathematics, astronomy, physics, chemistry, anatomy, physiology, botany, and so on. It is characteristic of this tradition that a great deal of attention is devoted to the conceptual development within the various fields of knowledge. In this way of writing history, the boundaries of the disciplines are often drawn in the same way as they can be found in recent textbooks, with the result that a history of electricity, for example, ignores the fact that for a long while no connection was made between lightning and the attractive power of amber. One does not learn from a history of this kind about the point in time at which the research field 'electricity' came into being and the reasons for which it did so.[4]

In Kuhn's view the other tradition, which regards science as a single entity, cannot be reproached for these omissions, yet the drawback to this tradition is that, generally speaking, it devotes little attention to the content of knowledge. It concentrates instead on the changing intellectual, ideological, and institutional context of science. It can contribute little to a clarification of the relationship between the 'metascientific environment', on the one hand, and the development of specific scientific theories and experimentation, on the other, since the social or philosophical factors that promote the development of a particular discipline in one period, obstruct it in another, and in any particular period some factors operate to the advantage of one science and to the detriment of another.[5]

What Kuhn suggests amounts to a middle course between the two alternatives. He proposes that science should not be seen as a whole; however, the present division into disciplines should no longer be taken for granted either. The historian's first task is to discover the 'natural divisions', in other words, the divisions between areas of knowledge (often institutionally determined) that were utilised in the period under discussion. A 'disciplinary map' of science in any particular period is a prior condition for an investigation of the complex influence exerted by extrascientific factors on the development of science. According to Kuhn, it is equally important for an investigation of this kind that one should have extensive knowledge of the conceptual development within the various disciplines.[6]

Kuhn's proposal can be understood as a plea for historicising the history of science, concentrating on the use of divisions between disciplines. His selecting this subject for his address arose from his aim of bridging the gulf between 'internal' and 'external' historical writing, which to a great extent coincides with the gap between the two historiographical traditions outlined.[7]

What 'natural division' within science does Kuhn suggest? He provides only a partial answer, confining his attention to the *physical* sciences, within which he indicates two groups of disciplines: the 'classical' and the 'Baconian' physical sciences. Astronomy, statics, and geometrical optics belong to the first grouping. These sciences had a tradition of research even in Greek Antiquity, hence the name classical. The second grouping includes the study of magnetism, electricity, and heat. Study of these subjects only developed into important areas for research after 1600; before Bacon's time they had been subject to little investigation. Kuhn links the Baconian sciences with what he calls the experimental tradition, and he connects the classical sciences with 'the mathematical tradition'.[8]

There is little that is new up to this point. Historicising organisational concepts is professed by every self-respecting historian of science, and even Kuhn's concrete elaboration of this, as applied to disciplinary divisions, follows a path trodden by others before him. Many writers have in one way or another drawn attention to the existence of these two traditions and to the fact that the two groups of physical sciences linked to these traditions developed in different ways.[9] What, then, is new or important in Kuhn's scheme? Its first merit is that it renders general and explicit what others have used only in part or implicitly. A second and crucial difference is that Kuhn shows how useful a conscious use of his scheme can be in the elucidation of two historiographical problems, both famous and notorious. These concern the origins of modern science in the sixteenth and seventeenth centuries and the rise of modern physics as a discipline in the nineteenth century. The two developments are often referred to as the first and second scientific revolutions.[10] A third feature of Kuhn's approach has already been mentioned. This is the aim of bridging the opposition between internal and external historiography.

What resonances has Kuhn's scheme evoked? Some of Kuhn's students who became acquainted with his ideas long before 1975 used them in setting up their own historical research. Heilbron's monumental study on seventeenth- and eighteenth-century electricity is a good example of Kuhn's influence in this respect.[11] Others, with a less direct relationship to Kuhn, have also utilised his scheme. R. W. Home, for instance, used it in order to clarify a historiographical problem that was different from the two problems Kuhn himself had raised. Home set himself the task of showing why so little value was initially placed on Aepinus' book on electricity and magnetism (*Tentamen theoriae electricitatis et magnetismi*, 1759), which is now recognised as a classic in the field. Home's solution is a simple one: The book had no public in mid–eighteenth century. It had found no serious readers because Aepinus' approach was unusual for its time: a systematic procedure entailing elementary but strictly applied mathematics and consequences based on calculations that were confirmed by experiments. Home believes that, on the one hand, the mathematics was too simple to be of interest to those working within the existing mathematical tradition, and on the other hand, the proportion of mathematics in the book was too great for the experimenters; even if they were not hostile to the use of mathematics in experimental physics, they found the book incomprehensible. Aepinus employed a method that combined elements from both traditions. His book only found a group of like-minded readers at the end of the eighteenth century and the beginning of the nineteenth.[12] The explanatory function of Kuhn's scheme was also recognised in general surveys. Hall, for example, made intensive use of this in rewriting his book on early modern science, and Schuster used it in giving his succinct but interesting account of the Scientific Revolution.[13] Finally, a recent book-length survey of the literature on the Scientific Revolution has a full treatment of Kuhn's proposal.[14]

2. An addition to the scheme: The natural philosophical tradition

A thesis as useful as Kuhn's clearly demands to be tested against the results of personal research. Does his scheme shed light on the history of physical optics?

The obvious first step is to examine the way in which Kuhn's scheme has been applied to the history of optics, up to the present day. In an article published in 1961, in which Kuhn uses his scheme for the first time (with somewhat different terms), optics forms an important example. Kuhn regards optics, that is, geometrical optics, as a classical science and part of the old mathematical tradition. However, in his view, early modern optics also contained a part that developed along experimental or Baconian lines. This Baconian part came into being in the second half of the seventeenth century and comprised experiments on interference, diffraction, and polarisation (Kuhn uses the present-day terminology). This area of optics remained separate from its mathematical part and was included into a mathematical theory only at the beginning of the nineteenth century. According to Kuhn, this last development formed an important part of the mathematisation of the Baconian physical sciences. In the 1975 essay his statements on optics are distributed throughout his text, but taken together they form the same picture as that given fourteen years earlier.[15]

Few others have written about the place of optics within Kuhn's scheme; Cantor is one of them. In his book on optics in Britain and Ireland between 1704 and 1840, Cantor holds that 'it is not clear whether physical, as opposed to geometrical, optics fits neatly into either category', that is, the categories of classical and Baconian sciences. For Cantor this signifies a 'difficulty with Kuhn's thesis in respect to the eighteenth century', and he maintains that 'Kuhn's scheme requires some modification'.[16] He does not say what kind of alteration is required. Nonetheless, I agree with Cantor's conclusion that physical optics does pose a problem for Kuhn's scheme. Indeed, these difficulties are a symptom of a fundamental shortcoming in Kuhn's thesis.

In my view, Kuhn's two-fold division should be replaced by a three-fold one.[17] The three traditions I would like to distinguish are the mathematical, the natural philosophical, and the experimental. There are three concomitant groups of physical sciences, the first being mathematics in its widest sense, including 'mixed', or applied, mathematics, which also comprises the mathematical areas of science. The second group is what, for lack of a better term, I call the theoretical portions of natural philosophy, or theoretical natural philosophy (there is a conscious decision here not to opt for 'theoretical physical science', because this term refers to later developments). Third and last are the experimental parts of natural philosophy, or experimental natural philosophy.

The main reason for introducing a third tradition in the development of early modern physical science is historical continuity. In medieval times important subjects that nowadays are treated in the physical sciences were part of Aristotelian natural philosophy, which in its turn was closely linked to metaphysics. The goal of natural philosophy was to provide a complete account of nature. This account was considered well established because it was based on metaphysical and empirical certainties. This *ideal* of a complete, certain, and partly a priori picture of the natural world did not die when the *contents* of Aristotelian natural philosophy were rejected in the Scientific Revolution.

Descartes' *Principia philosophiae* is a case in point, and a crucial one. His aim was to create and justify an alternative for Aristotle's philosophy. His book was literally meant to replace the Aristotelian corpus in education. Though Descartes' metaphysics and natural philosophy were at odds with Aristotelian orthodoxy, his goals and methods were much less so. He continued to strive for a complete as well as certain picture of the world. Basic concepts and laws were founded a priori in his metaphysics. Of course, apart from the glaring differences in content, there were also some differences in goals and method. Contrary to Aristotle, Descartes aimed for a quantitative, mathematical view of the natural world. In addition, he gave a role to experimentation as a trustworthy instrument of knowledge. However, Descartes' system of the world was not mathematised in the sense that Newton's or Laplace's would be: There was no set of mathematical equations from which the movements of the planets were derived in a quantitative way. Descartes' *vortices* were not mathematised, and they provided an essentially nonquantitative explanation of planetary movement. Moreover, Descartes' use of experimental data, for example, in his discussion of magnetism, was much in the same vein: There were no mathematical equations at all. The net result was a nonquantitative, visualisable, and complete explanation of the natural world in terms of a priori established concepts. So, what was new *in principle* in Descartes' approach did not *in fact* transform the epistemological claims and methods of natural philosophy. Therefore, Descartes' *Principia philosophiae* falls squarely in the natural philosophical tradition.

The *Principia philosophiae* did not gain the widespread use Descartes would have liked, but mechanistic natural philosophy, Cartesian or otherwise, did. After some time many schools and universities used mechanistic textbooks on natural philosophy. It is difficult to find a place for these texts and their contents in either one of Kuhn's categories. They constitute a rather well-defined and separate stream: natural philosophy. Of course, this changed gradually as time went on. Clarke's edition of Rohault's Cartesian textbook was 'undermined' by footnotes that grew larger and larger with every new edition. And this particular example was parallelled by less dramatic but similar changes elsewhere. Somewhat later, at many places 'Newtonian' texts entered the schools. One might say that the natural philosophical *tradition*, as a separate disciplinary unit, came to an end somewhere during the eighteenth century. The picture delineated here gives a first account of what in Kuhn's account is, at best, to be described as the mysterious disappearance of the natural philosophical tradition in the early seventeenth century (see Section 4, below). It did not disappear as a disciplinary unit for well over a century, and its disappearance (as a separate disciplinary unit) and heritage (goals, concepts, methods) are not mysteries. Both processes can be traced through the twists and turns of eighteenth- and early nineteenth-century physical science, as well as through the concomitant developments in philosophy.

In the German lands the natural philosophical tradition has been especially visible as a separate disciplinary unit, well into the second half of the eighteenth century.[18] The special visibility and longevity of the natural philosophical tradition in Germany is closely connected to the combined influence of Leibniz, Wolff, and the Prussian Academy. The goals and methods, if not the contents, of much of Leibniz' theorizing about the natural world fall into the Aristotelian and mechanistic traditions of natural philosophy. Leibniz' ideas were very influential in eighteenth-century Germany. Wolff often acted as an intermediary. He systematised Leibniz' thoughts in metaphysics and natural philosophy and used a classification of the sciences that had close parallels with the three-fold division advocated here. Wolff's textbooks followed this scheme, as we saw earlier. He wrote a book on the mathematical sciences, in which the mathematical parts of physical science found their place, including Newton's mechanics. A second and separate volume, in its German version commonly known as *Deutsche Physik*, covered the theoretical parts of natural philosophy, including the basic ideas of Leibniz and Descartes. A third volume was exclusively devoted to experiments; the German version was commonly designated as *die Experimentalphysik*. It contained experiments by Galileo, Mariotte, Boyle, Newton, and van Musschenbroek.[19] The same three-fold disciplinary division was used when the Prussian Academy was reorganised in 1746. It was structured in three 'classes': those for 'philosophie spéculative', 'mathématiques' and 'philosophie expérimentale'.[20]

What is the relevance of the amended Kuhnian scheme, the three-fold division, for the history of early modern optics? Put in a nutshell, it is that eighteenth-century physical optics can be characterised as a science torn among the three traditions mentioned and that this characterisation helps us to understand some of its patterns of development.

What later became united in one field, physical optics, was represented in each of the three groups of sciences. Because of the disciplinary barriers between the three traditions and the lack of experience by those who tried to combine two or three of the traditions, the three 'parts' of physical optics to a great extent developed separately from one another and embraced different methods and objectives. In order to provide an impression of what I mean by the three traditions in optical science, I shall illustrate these terms by discussing a number of well-known developments between 1690 and 1830.

Huygens' pulse theory of light is a good example of a theory that, although limited in its range, achieved a far-reaching degree of synthesis among the three sections of physical optics in the area it covered. The medium conception of light, and Huygens' principle, form the *natural philosophical* heart of the theory. Huygens' principle can be translated into geometrical terms and thus underlies the familiar laws of *mathematical*, or geometrical, optics, such as the laws of reflection and refraction. The *experiments* on double refraction can be explained by means of an additional physical hypothesis on two kinds of propagation in Iceland crystal. We are so accustomed to the methodological pattern in Huygens' optical research that it comes as a surprise to realise its rarity, if not uniqueness, in the early modern period. A synthesis of natural philosophical 'speculation', mathematical techniques, and careful experimentation is perhaps without parallel in physical optics before the end of the eighteenth century.[21]

Newton's achievements in the area of physical optics are no less important than those of Huygens, but they are of a different kind. His strength lies within the experimental area of physical optics. His *Opticks* is presented as a systematic elaboration of *experiments*. His aim was to develop a *mathematical* theory of colours on the basis of these experiments, but he failed, apart from his theory of fits.[22] His *natural philosophical* ideas on light are provided in separate Queries. Moreover, these natural philosophical ideas remain at the suggestion stage, being neither critically sifted and systematised nor *mathematised*. An isolated example of mathematising his natural philosophical model of light is the deduction of the law of refraction on the basis of the assumption that the denser medium exerts a net attractive force on a light particle.

If Newton has not significantly integrated the natural philosophical side of physical optics with the two other parts, in Euler's case there is an incomplete integration of his theory, which is both natural philosophical and mathematical, with the experimental part of physical optics. This statement requires some explanation, since Euler has been described above as the one who finally integrated Newton's colour experiments into the medium tradition. Why do I believe, nonetheless, that the experimental side of Euler's physical optics reveals a lack of cohesion with his theory of light?

The reason is that Euler's incorporation of the new theory of colours into his medium theory of light was carried out almost entirely on the level of natural philosophy and mathematics. Euler took over Newton's building and used his own natural philosophical and mathematical principles as its foundation. Granted, he did adapt Newton's edifice in places, and he provided new physical explanations for known experimental results on dispersion and Newton's rings, but – this is the crucial feature – he did so without carrying out *new* experiments in order to establish whether the empirical consequences of his adaptation and incorporation were correct.

For Euler the methodological process that characterises Huygens' optical research failed to get off the ground. This process can be described as the road from theory, or natural philosophical model, to mathematically derived results and experimental testing, or, to put it the other way round, from remarkable experimental facts to necessary adaptations of the theory, or its mathematical formulation. Expressed even more precisely, the process demands the combination of both methods and thus continual interaction between theory, mathematics, and experiment. It is this approach that characterises at least part of Huygens' *Traité*, and it resurfaced in physical optics only at the end of the eighteenth century.[23]

None of this means that Newton, Euler, or others would have attained no significant results. Both the emission theorists (Malus, Biot, Laplace) and wave theorists (Young, Fresnel) at the beginning of the nineteenth century pursued the work of the 'Newtonians', Euler, and others. However, the French researchers in particular introduced a methodological renewal in respect to Newton's and Euler's work and that of the whole field of eighteenth-century physical optics. Once again, they applied the method that Huygens had already exhibited. Once again, they integrated natural philosophical theory, mathematics, and experiment in such a way that equal attention was given to all three of these elements and they were able to cross-fertilise one another.

This, then, is the general argument for, and the specific illustration of, my three categories. I shall now give several examples of the way in which the three-fold division can provide historiographical clarification of developments in eighteenth-century optics as they were described in the foregoing chapters. The discussion of these examples at the same time presents a summary and a deeper examination of some of the results they contain.

3. The three-fold division and eighteenth-century optics

a. The reception of Newton's theory of colours

Newton's theory of colours, with its principle thesis that white light is a heterogeneous mixture of the colours of the rainbow, was accepted with little difficulty in Germany shortly after German scholars had become acquainted with it via the *Opticks* (1704) and the *Optice* (1706). Wolff and Scheuchzer were the chief disseminators of the new theory of colours. They made a clear distinction between the natural philosophical question concerning the nature or physical structure of light, on the one hand, and, on the other, what was to them the purely empirical question of whether the new theory of colours offered a correct description of colour phenomena. Where the latter point is concerned, Newton's experiments and argumentation had convinced them, which is not to say that they also adopted Newton's conception of the nature of light, his emission hypothesis. Both Wolff and Scheuchzer put forward a medium conception of light.

It is perhaps natural to assume that the experimental and the natural philosophical components of physical optics form a cohesive whole, but from this example it is clear that this does not always hold good. In Wolff's case optics was neatly divided over his three textbooks. Geometrical optics, including the law of refraction (with no physical explanation), was discussed in his textbook on 'mathematics', his volume on 'theoretical natural philosophy' included the medium conception of light, and his work on 'experimental natural philosophy' contained an experimental proof of Newton's theory of colours.

Analysis of the reception of Newton's theory of colours in France once again shows the illumination to be obtained from the three-fold division. In France, too, it was an influential medium theorist, Malebranche, who embraced the new theory of colours after reading Newton's book on optics (the Latin edition of 1706). His judgement of Newton's work, cited earlier, illustrates the way in which the three-fold division was actually used in that period:

> Quoique Mr. Newton ne soit point *physicien*, son livre est tres curieux et tres utile à ceux qui ont de bons principes de physique, il est d'ailleurs excellent *geometre*. Tout ce que je pense des proprietez de la lumiere s'ajuste à toutes ses *experiences*. (Emphasis added)[24]

Here it is difficult not to see the reflection of the three-fold division among sciences. Newton is portrayed as an authority on experimental natural philosophy and mathematics but as an ignoramus in the field of 'physics', that is, in matters of theoretical natural philosophy. Malebranche considered himself to be the real initiate in this last area.

b. The reception of Euler's wave theory in Germany

Euler's 'Nova theoria', published in 1746, was favourably received in Germany. The majority of German scholars accepted a medium hypothesis for more than thirty years. When we examine this fact in greater detail, it is useful to distinguish between two different kinds of literature: the textbooks and encyclopaedias, on the one hand, and, on the other, the research reports, such as journal articles, monographs, and on one or two occasions a passage in a textbook or encyclopedia falling into the category of research reports. In spite of the differences between the two kinds of literature, the reception of Euler's theory followed the same path in both of them. This characteristic pattern can easily be understood in terms of the division into three scientific traditions.

If a *teacher* accepted Euler's theory of light, that acceptance was registered in a textbook on theoretical natural philosophy (*Naturlehre*) in which the theory was regarded as the answer to the fundamental enquiry into the nature of light. Because students attending the lectures on physics did not know enough mathematics, the teacher omitted the mathematical part of Euler's theory.[25] Was that part of the theory dealt with in textbooks on 'mixed', or applied, mathematics? The answer to this is no, because questions concerning the physical explanation of natural phenomena did not belong in the field of mathematics, but in the realm of natural philosophy. Since the mathematical part of Euler's wave theory could not be treated separately from the physical nature of light (something that was actually possible in the case of Newton's theory of colours), a teacher of mathematics would have to take sides in a question that concerned natural philosophy. Apart from one exception, who actually confirms the general pattern by apologising for breaking the code of proper behaviour,[26] nobody wanted to cross the border between the two fields of knowledge.

Euler's theory was not only split into two sections in this way; it was also decapitated. The mathematical part of his theory fell between two stools and thus attracted no attention at all. In contrast, his natural philosophical point of view was treated and accepted in the majority of natural philosophical textbooks, albeit sometimes in the less specific form – the medium rather than the wave conception.

We saw that the same sad fate befell Huygens' theory. In Huygens' case the consequences of the rigorous division between the separate parts of his optical theory were even more dramatic. Huygens' 'crucial experiment' on double refraction and its mathematical theoretical explanation were chopped into three chunks. The mathematical aspect was too difficult or too much bound to natural philosophy and therefore omitted in natural philosophical or mathematical textbooks, respectively. Huygens' theoretical position, particularly his pulse-front principle, was certainly mentioned in natural philosophical textbooks, but the study of this was left to the students themselves, or else a distorted version of the theory was reproduced. The experimental part of Huygens' *Traité de la lumiere*, the empirical laws of double refraction, was found even less frequently in textbooks. If the phenomenon of double refraction was mentioned at all, it was presented as an experimental curiosity and described in a qualitative manner. In this way a highlight in the seventeenth-century optical literature – the only work in which the three components of mathematics, natural philosophy, and experimentation were successfully balanced and united into one harmonious unity – was disfigured to the point of being unrecognisable by the divisions made between the three groups of physical sciences.

Let us now return to Euler. How was his theory of light received in the *research literature*? Mutatis mutandis, we find the same pattern here as in the educational literature. In research reports, as well, absolutely no attention was paid to the testing or improvement of the mathematical side of the theory. The Germans who accepted (or indeed rejected) Euler's theory, concentrated on the basic idea that light consisted of a motion within a subtle medium; even the periodic nature of the wave motion was hardly ever discussed. Consequently the research was of a qualitatively experimental kind or consisted in the repetition of, or finding variations of, Euler's standard arguments, with no purposeful experimentation or theoretical work of any depth. Thus, at best these people integrated two of the three traditions, that of theoretical natural philosophy and of experimental philosophy. Nobody in Germany developed Euler's theory further or challenged it by integrating all three traditions or by working at least on the borders of mathematics and natural philosophical theory, where Euler himself had been active.

The absence of anyone to combine the three traditions within physical optics accords with Home's conclusions on the reception of Aepinus' theories on electricity and magnetism. As we have seen, Home regarded these conclusions as illustrating the productiveness of Kuhn's two-fold division. There is, however, a difference between the reception of Euler's theory of light and that accorded to Aepinus' theories, which makes it clear why a three-fold division is preferable to a two-fold split. After all, Aepinus' theory was ignored or rejected, yet Euler's wave theory was accepted by one of the three groups, the natural philosophical one.

One more word about a general question that follows naturally upon what we have just been discussing: What prevented a more complete mathematisation of physical optics in the eighteenth century? Where medium theories are concerned the answer consists of a generalisation of what has been said about Euler and Huygens. Simplifying and disregarding one or two exceptions, this can be formulated in the following way. In general there was insufficient contact between the three sections of physical optics, as there was not much contact between the three concomitant traditions in general. The medium theories were torn into separate pieces by the three traditions and subsequently reduced to rubble.

Something similar obtains for the emission theories. Even though in this case there was, at first sight, a greater degree of contact between the mathematical, theoretical and experimental parts, upon closer examination this contact was clearly not enough to secure change in the direction of mathematisation. After all, the integrating activities consisted mainly of the sorting and systematisation of remarks and suggestions, particularly Newton's, distributed over several sources, and were intended as teaching material. Just as in Euler's case, there was a lack of active exchange between theory, mathematics, and experiment. The mathematisation of the emission tradition gained significant momentum only in the early nineteenth century, when the emission hypothesis became part of a deliberately organised programme of research, integrating the three aspects of modern physics. This development makes it clear that the problem of the obstacles to the mathematisation of physical optics is actually expressed in the wrong terms. The very word *mathematisation* suggests that the only requirement was for increasing application of mathematics, but the real stumbling block for the further development of optics was the lack of an integrated method from which mathematisation could follow as one of its consequences.

c. The chemical turn in physical optics

As we have seen, the German debate on the nature of light followed a remarkable course. For a long time there was no final decision, although the medium tradition enjoyed the ascendency for many years. In this period the two factions advanced variations on the standard arguments or made vain attempts to reach a decision by performing or suggesting new experiments. Meanwhile the medium tradition retained its dominant position.

This picture was completely altered within the short time span between 1789 and 1795. The ascendency of the medium tradition gave way to the marked dominance of the emission tradition. This radical turn was accompanied by a new argument in support of the emission theories, the 'chemical argument' mentioned above. This reasoning, supported by the growing prestige of chemistry and the chemists themselves in the period we are examining, rapidly brought the lingering discussion to an end. For many people the chemical effects of light (important examples of this being the blackening of silver salts and the production of oxygen by green leaves) constituted the indisputable proof that light was an active constituent in chemical processes. In their view chemistry thereby demonstrated that light was material and that consequently the emission hypothesis was correct.

This episode in the German optical debate was a remarkable phenomenon from the point of view of theory change. After all, physical optics was a partly mathematised field, and one would perhaps expect this mathematisation to be stifled occasionally but never actually reduced. Yet the chemical argument is of the qualitative kind and tends towards demathematisation, since the theoretical ties with mechanics are loosened, if not cut. The deviant character of the chemical turn within physical optics is all the more striking if we compare it with developments in electrostatics in the same period. In Kuhn's division this is a Baconian field, the development of which is, according to Heilbron, typical of the Baconian sciences in the eighteenth century.

Quantification and mathematisation reigned supreme in this area in the last decades of that century. Heilbron believes electrostatics at this time to be typified by a decreasing influence from traditional natural philosophy ('speculation') and by an increasing 'instrumentalism'.[27] In contrast, the development of German optics showed increasing influence from a chemical, 'old-fashioned' way of devising theories and the accompanying demathematisation of a partially mathematised field. Thus, a historical analysis with the aid of two scientific traditions and the linked thesis that the physical sciences within the experimental tradition were gradually mathematised is not sufficient in this case. It is clear that a third tradition, the natural philosophical, is no less able to determine the manner in which the development of a physical science takes place.

This particular example would lose much (although not all) of its historiographical importance if it were possible to maintain that the chemical turn in optics was an exclusively German affair with no international significance. However, the chemical turn in optics seems to have taken place in France and Britain, too. In 1986, when the first (Dutch) edition of this book was published, there were some indications in the literature to that effect. Meanwhile, Shapiro has provided ample evidence that physical optics in France and Britain achieved an important overlap with chemistry. As was shown above, the special circumstances in Germany, where Euler's influence had ensured the dominance of the medium tradition, were responsible for the fact that the link with chemistry was revealed there in a remarkably clear way. Shapiro also suggests that the chemical turn in optics implies a fundamental change in our view of the physical sciences around the turn of the century. Employing analytical terms that are consistent with the three-fold division used here, he concludes that we should 'reevaluate our traditional emphasis on the mathematical and experimental aspects of physical optics in this period and somehow relate it to the apparently more widespread chemical view of light'.[28]

4. On early modern science

In conclusion, I would like to make two observations. In the first place there are insufficient grounds for altering Kuhn's scheme in any fundamental way, as long as it has not been demonstrated that the two historical developments Kuhn selected for discussion gain any clarification from the three-fold division. These historical developments are the two scientific revolutions. The first of these, taking place in the sixteenth and seventeenth centuries, consisted (in Kuhn's view) of a conceptual transformation of the classical, or mathematical, sciences and the emergence of a group of new sciences, the Baconian, or experimental physical, sciences. In my opinion this analysis does no justice to the natural philosophical aspects of the work carried out by such important figures as Kepler, Descartes, Huygens, Leibniz, and Newton, to say nothing of the university professors teaching natural philosophy. It is better not to portray the second scientific revolution (the emergence of modern physics as a discipline beginning in France shortly before 1800 and later spreading to other countries) in the way Kuhn does, that is, as a mathematisation of the Baconian physical sciences. The process that actually occurred can more satisfactorily be comprehended as a synthesis involving three traditions, all of which underwent alteration in the course of this process.

These are tentative suggestions that do not claim or aim to put my main thesis – the changing of Kuhn's scheme – beyond doubt. The examples taken from eighteenth-century optics are mainly illustrative rather than argumentative.[29] Nevertheless, I hope to have found acceptance for my second thesis, that eighteenth-century optics was divided between the three traditions and therefore followed lines of development different from those one might have expected on the basis of Kuhn's scheme.

The second and last observation once more concerns my main thesis, that Kuhn's two traditions should be replaced by three traditions. This observation is an attempt to make explicit the core of my proposal, which concerns the place of natural philosophy in early modern science.

There is no difference of opinion between me, on the one hand, and Kuhn and Heilbron, on the other, about the fact that the historiography of science has laid too much emphasis on general cosmology, fundamental questions concerning natural philosophy, and so on. Kuhn's stressing of the disciplinary aspects of science as he himself, Heilbron, and others applied them is a justified corrective to the historiographical tendency to allot a universal, all-determining influence to natural philosophical aspects. However, in my opinion they have gone too far in this direction, thereby rendering these aspects virtually invisible.

The essence of my proposal is perhaps a difference of opinion over the concept of science – and of early modern science in particular. This can be illustrated by some passages from Kuhn's 1975 essay, at the end of which he posits that the separation of the mathematical and experimental sciences survives in the late nineteenth-century division, within physics, between theoretical and experimental physics. Here Kuhn notes that present-day physical theories are 'intrinsically mathematical'.[30] In my conception, theoretical natural philosophy (we would now be talking of the conceptual aspects of a physical theory) is just as much a part of physics as are mathematics and experimentation. Furthermore, however one wants to define twentieth-century physics, it is much more difficult to leave out the natural philosophical component from early modern physics and science. 'To leave out' is rather too strong a way of putting it. In his paper Kuhn occasionally brings natural philosophy onto the stage, and he even talks explicitly about 'the ancient and medieval philosophical tradition'. Further, when he mentions seventeenth-century atomic or corpuscular natural philosophy he describes it as a 'set of metaphysical commitments' leading to sterile 'metaphysical theory'.[31] Thus, Kuhn is not blind to the natural philosophical tradition, but he appears to deny its importance because in his view it is a metaphysical tradition rather than a scientific one. If this description of Kuhn's position is correct[32], then this is a debatable but defensible vision of twentieth-century physics. Yet as a historical image of early modern science, this point of view does reveal an anachronistic tendency. Kuhn appears to stress the 'modern' in early modern science, while I would rather emphasise the 'early'.

NOTES

1. Introduction

1. For a description of the standard view, centring on the situation in Britain, see Cantor, 'Historiography' and id., *Optics*, 1–15. Recent studies on aspects of eighteenth-century optics: Cantor, *Optics*, and Shapiro, *Fits*; see also Steffens, *Development*, and Pav, *Optics*. For an older work on eighteenth-century science, see Wolf, *History*; Hankins, *Science*, gives an overview based on recent secondary literature but gives no consideration to physical optics.
2. Compare Cantor, *Optics*, 25–146. Cantor barely discusses the development of the medium theory in the first half of the eighteenth century, and he gives Euler's wave theory of light only a short summary. Cantor also devotes little attention to non-'Newtonian' emission conceptions and to the arguments for and against the different conceptions. All these differences from my study derive from the fact that Cantor confines his attention to developments in Britain and Ireland.
3. For the history of optics, see, for instance, Ronchi, *Histoire*.
4. Hankins, 'Reception', especially 59–60; Truesdell, *Essays*, 116–17.
5. Wolf, *History*, 161; Weinmann, *Natur*, 113. Ronchi, *Histoire*, 215–17, restricts the thesis to 'the public at large'.
6. There is an interesting exchange of letters and articles in a Dutch newspaper and the journal of the Dutch Society of Physicists on this claim. H. de Lang, a physicist, criticising my exposition of the fate of Huygens' theory in the eighteenth century, ignored the fact that it was not I who rejected Huygens' theory, but many contemporaries for about a century after its publication, and he interpreted my account as my *present-day* evaluation of Huygens' theory, assuming that I would share his presentist perspective on the history of optics. He also claimed that 'Huygens' theory' was very well able to explain rectilinear propagation, even without using the concept of periodicity. This is a fine example of historical condensation, since de Lang, for instance, uses the concept of destructive interference of waves in his account; however, this concept is not to be found in Huygens' *Traité*. Hakfoort, 'Schaduwzijde'; 'Naschrift'; 'Waarom'; de Lang, 'Huygens'; 'Lichtvoortplantingstheorie'; 'Weerwoord'; 'Originator'.
7. The term *wave* includes the concept of pulse, both in present-day technical usage and in many eighteenth-century texts. However, as we will see in Chapter 4, Section 3b, the change from nonperiodic pulses to periodic waves was an essen-

tial step in Euler's theory of light. Moreover, reserving the term *waves* for periodic waves, reminds us that Huygens' 'waves' were not periodic (Dijksterhuis, *Mechanisering*, 507).

8. Planck, 'Wesen'; Wiener, 'Wettstreit'; Rosenfeld, 'Premier conflit', with its opening sentence: 'Quoi de plus passionnant, dans l'histoire de la Science, que cette lutte où s'affrontent, dès leur naissance, les deux conceptions, continue et discontinue, de la nature de la lumière?' Compare Cantor, *Optics*, 4–6, on Whewell's influential historiography of optics in his *History of the inductive sciences* (1837). Whewell was also writing at a time in which there was a debate between 'the' emission theory and 'the' wave theory. His historiography of optics seems intended to support the position of 'the' wave theory in the debate.
9. Weinmann, *Natur*, 59.
10. Ibid., 58–67.
11. Hoppe, *Geschichte*, 6.
12. Nagel, *Structure*, 83.

2. The debate on colours, 1672–1720

1. Aristotle, 'De sensu', chaps. 3 and 6, 445b3–446a22; id., *Meteorologica*, 374b7–375a30. Westfall, 'Development', 342.
2. Mac Lean, 'De kleurenleer', 211.
3. Gilson, *Études*.
4. Descartes, *Principia*, 108–16. See also Chapter 3, Section 2a.
5. Descartes, *Dioptrique*, 89–92; id., *Meteores*, 331–5.
6. Westfall, 'Development', 344–8 discussing Grimaldi, Boyle, and Hooke.
7. Newton, *Correspondence*, vol. 1, 92–107.
8. For the way in which Newton's theory of colours was actually developed, see McGuire and Tamny, *Certain*, 241–74; Newton, ed. Shapiro, *Optical papers*, vol. 1, 1–25, and the literature given there.
9. Sabra, *Theories*, 240–1.
10. Newton, *Correspondence*, vol. 1, 92.
11. Descartes, *Dioptrique*, 100–2.
12. Newton, *Correspondence*, vol. 1, 93.
13. Ibid., 166. Compare Newton, *Papers*, 101.
14. Newton, *Correspondence*, vol. 1, 94–5, quotation 95 (emphasis in the original text).
15. Ibid., 97, 98 (emphasis in the original text).
16. Ibid., 100 (emphasis in the original text).
17. Ibid., 105–6. Cohen, 'Versions', provides a facsimile of the text. According to Hall, 'Études', 51–4 the text was printed in 1676 or 1677 but never published. It was thus probably unknown to Newton's contemporaries. Compare Sabra, *Theories*, 244–5, for another interpretation of the quotation.
18. Newton, *Correspondence*, vol. 1, 205–6, 207.
19. Ibid., 113–14, 131.
20. Ibid., 110, 157.
21. Westfall, 'Critics', especially 51.
22. Newton, *Correspondence*, vol. 1, 235–6, 255–6, quotation 236.
23. Ibid., 142, 173–4, 264.
24. Ibid., 175. Sabra, 'Vibrations', 267–8; Blay, 'Clarification', 216–17.

25. Mariotte, *De la nature des couleurs* (Paris, 1681). Edition used: *Oeuvres*, vol. 1, 195–320. The passage from 1681 on Newton's *experimentum crucis* is cited in Lohne, 'Experimentum', 186; the *Oeuvres* text does not diverge from this.
26. Mariotte, *Oeuvres*, vol. 1, quotations 226, 228.
27. It was assumed later that Mariotte had not sufficiently separated the colours. Extra care has to be paid to this because the sun is not a mathematical point, and light from the sun's surroundings can also easily end up in the spectrum; Desaguliers, 'An account of some experiments of light', 434–5.
28. Mariotte, *Oeuvres*, vol. 1, 228 (emphasis in the original).
29. Guerlac, *Newton*, 98–9.
30. Ibid., 102–3. Cohen, 'Sloane'; id., *Newtonian revolution*, 146 talks of ten sessions.
31. Guerlac, *Newton*, chap. 5, especially 129–31, 138–9, 162–3.
32. Leibniz to Oldenburg, 8 March 1673; Leibniz, *Sämtliche Schriften*, vol. III.1, 43–4.
33. Knorr, *Dissertatio*. Leibniz to Huygens, 26 April 1694; Huygens, *Oeuvres*, vol. 10, 601–2. According to the title page of the *Dissertatio*, Knorr was chairman and the dissertation was defended by J. J. Hartman. I am following Leibniz and the editors of the Huygens correspondence, who regard Knorr as the real author.
34. Knorr, *Dissertatio*, sect. 26–8.
35. Leibniz to Huygens, 26 April 1694; Huygens, *Oeuvres*, vol. 10, 602. Leibniz to de Breyrie, 8 May 1694; Leibniz, *Opera*, vol. 3, 662.
36. After mentioning Mariotte's objections, Leibniz says, 'Sed Newtono, qui id diu et pene unum agit, fidere malim, donec ad experimentum venire liceat'; Leibniz, *Nachgelassene Schriften*, 106.
37. Leibniz to J. Lelong, 9 April 1708; Malebranche, *Oeuvres*, vol. 19, 784. A few months before Leibniz had asked, through Lelong, for Malebranche's opinion on colours, in the light of Newton's experiments and Mariotte's objections (Leibniz to J. Lelong, 13 December 1707; Malebranche, *Oeuvres*, vol. 19, 768; an answer by Lelong to Leibniz is not known).
38. Wolff, Review of Newton, 59–64; attributed to Wolff in Ludovici, *Entwurff*, vol. 2, 186. Hall, *All*, 93, 206, suggests the review is by Leibniz or Otto Mencke, the editor of the *Acta*, but gives no evidence.
39. Wolff, *Anfangs-Gründe* (1710), vol. 3, 24–7.
40. Ibid., vol. 4, 460.
41. 'Objectiones, quas Viri docti tum in Gallia, tum in Anglia contra illam theoriam fecere, felicissime diluit Vir perspicacissimus *Newtonus*, quemadmodum ex Transactionibus Anglicanis . . . abunde constat: unde multi optant, ut mentem suam aperire dignetur de difficultate ab ingeniosissimo *Mariotto*, rerum naturalium (dum viveret) scrutatore indefesso, nec infelici, in Tractatu de Coloribus . . . contra eam mota'; 'Hac vero transmutatione admissa corruere theoriam *Newtonianam* . . . manifestum est. Assumsit autem *Mariottus* distantiam 30 pedum, ne quis exciperet, in minori distantia nondum factam esse plenariam radiorum heterogeneorum separationem. Nobis experimentum Mariotti tum demum videtur decisivum, si lumen coeruleum integrum in aliud mutatum fuisset'; Wolff, Review of Rohault, 447–8 (emphasis in original).
42. Priestley, *History*, vol. 1, 351; Pav, *Optics*, 138; Guerlac, *Newton*, 116.

43. Gerhardt, ed., *Briefwechsel*, 11, 94, 97, 100–4, 120–1, 177–8. See also Schlosser, *Rezensionstätigkeit*, especially 16–17.
44. Wolff, *Elementa*, vol. 2 (1715), 370. The reserve shown here towards Mariotte's experiments is even clearer than in the *Acta* of 1713: 'Enimvero cum non totum lumen violaceum in alios diversos colores abierit; id saltem inde colligitur, separationem radiorum diversis coloribus imbutorum in prima refractione non fuisse absolutam, minime autem nobis experimentum istud sufficere videtur ad theoriam illam [Newton's theory of colours] funditus evertendam.'
45. Desaguliers, 'An account of some experiments of light'.
46. Hall, 'Newton in France', 243–4, posits that the *Acta* regarded Mariotte's experiment as decisive. As can be read above, in reality almost the opposite was stated. (Compare Hall, *All*, 205–8, where pre-1714 German support for Newton's theory of colours is acknowledged.) Westfall, *Never at rest*, 794–5, says that the new theory of colours began to be spread over mainland Europe only after Desaguliers' repetition of Newton's experiment.
47. Wolff, 'Notanda'.
48. Wolff, *Allerhand*, vol. 2 (1728), 503–8. I have been unable to consult the first edition of 1722, but it is unlikely that the passage on Newton's *experimentum crucis* has been changed to any substantial extent. Id., *Würckungen der Natur*, 188–93, especially 191; 394, 400.
49. Krafft, 'Weg', 135–6, 139–40.
50. Scheuchzer, *Physica*. I have seen the editions of 1701, 1711, and 1729. There is an unaltered new edition from 1703 and a fifth, posthumous edition of 1745 (Fischer, *Scheuchzer*, 27; Steiger, 'Verzeichnis', 3, 4, 6, 14, 18).
51. Scheuchzer, *Physica* (1701), vol. 1, 123; (1711), 97.
52. Ibid. (1711), 107–30, quotation 107.
53. Kaschube, *Elementa*; Praefatio, [6]; 127, 165.
54. Weidler, *Experimentorum*. Weidler is called chairman on the title page, and D. G. Zwergius is mentioned as 'auctor'. Zedler's encyclopaedia refers to Weidler as the author; *Grosses Lexicon*, vol. 9 (1735), 230–1. Goethe, *Schriften*, vol. I.6, 346: In 1725 Johann Friedrich Wucherer named Weidler as the author. (In the eighteenth century it was customary to attribute a dissertation to the chairman.)
55. Weidler, *Experimentorum*, 5–15.
56. Ibid., 16–17.
57. Ibid., 21–22.
58. See also his comments on Newton's *experimentum crucis*; ibid., 28.
59. *Grosses Lexicon*, vol. 9 (1735), 230–1. Weidler, *Institutiones*, 244–5.
60. Rizzetti, 'De systemate', 130. This volume of the *Acta eruditorum; Supplementa* has '1724' on its title page. The journal was probably published in instalments. Because the previous volume has '1721' on the title page, Rizzetti's article cannot have appeared earlier than 1721, and in view of its location in the volume, it was probably published in 1722.
61. Newton, *Opticks* (1704), book 1, 15–16.
62. Rizzetti, 'De systemate', 131–2.
63. Richter, 'De iis', 226. Rizzetti's 'Excerpta e novi exemplari' and 'Excerpta ex epistola' were published directly following Richter's reaction. They contained

improvements on Rizzetti's 'De systemate' and an excerpt from a letter from Rizzetti to the Royal Society, respectively.

64. In 1728 Desaguliers mentioned a 'Paper' by Rizzetti that an Italian had shown him ('Optical experiments', 596–7). It is not clear whether this was the first *Acta* article (i.e., Rizzetti, 'De systemate'). At all events, Rizzetti's first *Acta* article is referred to in Desaguliers' published repetition of the experiment in 1722 (Desaguliers, 'Account of an optical experiment', 207). In 1728 Desaguliers also spoke of a letter from Rizzetti written in response to Desaguliers' repetition in 1722 ('Optical experiments', 596–7). This is probably Rizzetti, 'Excerpta ex epistola'.
65. Westfall, *Never at rest*, 796. Desaguliers, 'Optical experiments', 596.
66. Rizzetti, 'Super disquisitionem' and 'Ad responsionem'; Richter, 'Defensio'; Rizzetti, 'Responsio'; Richter, 'De praecedente', especially 493.
67. Rizzetti, *De luminis affectionibus specimen physico-mathematicum* (Treviso, 1727). I have not seen this work.
68. Desaguliers, 'Optical experiments', 598–600.
69. Ibid., 600–7, 607–29.
70. In 1741 Rizzetti continued the battle with his book *Saggio*.
71. The quotation is cited in Newton, *Correspondence*, vol. 7, xli. It derives from a previous version of Newton's published 'Remarks', especially 320–1.
72. Nevertheless, the conspiracy theory has been accepted by historians, as well. It even blinded one of them to simple facts. Pav, *Optics*, 40, writes about Rizzetti's first article (c. 1722; see note 60, above): 'Leibniz readily allowed the *Acta eruditorum* to be used as a sounding board'. In 1722, however, Leibniz had been dead for six years.

3. Theoretical traditions in physical optics, 1700–45

1. For the changes in the queries, see Koyré, 'Études'. Unfortunately, there is still no variorum edition of the *Opticks/Optice*.
2. Newton, *Opticks* (1704), book 3, 132–7.
3. Ibid., 132–4, qu. 1, 4, 5, 9.
4. Newton, *Optice* (1706), qu. 19, 306: 'congenitis & immutabilibus Radiorum proprietatibus'. Compare Newton, *Opticks* (1730), 361. Translations of Newton's Latin text are mine. Wherever possible, I have used the English translation in *Opticks* (1730).
5. He treats of double refraction in two new queries preceding the two 'critical' queries; Newton, *Optice* (1706), qu. 17, 18, 299–306.
6. Ibid., qu. 20, 307: 'Annon errantes sunt Hypotheses illae omnes, quibus Lumen in Pressu quodam, seu Motu per Medium fluidum propagato, consistere fingitur?' Compare Newton, *Opticks* (1730), 362.
7. Not all of these are always noted or mentioned; Rosenfeld, 'Premier conflit', 118–20, mentions three, for example.
8. Newton, *Optice* (1706), qu. 20, 307.
9. Huygens, *Traité*, 58–9, 91. Newton, *Optice* (1706), qu. 20, 308–9.
10. Newton, *Optice* (1706), qu. 18, 304–6.
11. Ibid., qu. 20, 307.

12. Newton, *Correspondence*, vol. 1, 175.
13. Id., *Principia* (1687), vol. 1, 510–14.
14. Id., *Optice* (1706), qu. 20, 307–8.
15. Ibid., 237–45, especially 239: 'Qui hoc in animum suum inducere non possunt, ut quicquam novi aut recens inventi Accipiant, quod nequeant continuo Hypothesi aliqua explicare'. Compare Newton, *Opticks* (1730), 280.
16. Newton, *Optice* (1706), qu. 20, 309–10. In id., *Opticks* (1730), 364, these three objections are replaced by one other: How can two ethers affect one another without spreading or confusing one another's movements?
17. Newton, *Optice* (1706), qu. 20, 310–15, especially 310: 'sensibilis materiae'; compare id., *Opticks* (1730), 368; id., *Principia*, vol. 1, 509: 'si forte vapores longe tenuissimos & trajectos lucis radios excipias'.
18. Cantor, *Optics*, 84–6; Buchwald, 'Investigations', 335; id., *Rise*, 8–19; see also Section 1c below. Shapiro, *Fits*, 203–7.
19. Newton, *Optice* (1706), qu. 21, 315: 'Annon Radii Luminis exigua sunt Corpuscula, e corporibus lucentibus emissa . . . ?' Compare id., *Opticks* (1730), 370; here the text reads 'very small Bodies', but *exiguus* means 'small', not 'very small'.
20. Newton, *Optice* (1706), qu. 21, 317: 'tenebricosissimum & languidissimum colorum'. Compare id., *Opticks* (1730), 372.
21. Newton, *Optice* (1706), qu. 21, 317–18: 'alia aliqua Vi'. Compare id., *Opticks* (1730), 372.
22. Newton, *Optice* (1706), qu. 21, 316: 'actio et reactio'. Compare id., *Opticks* (1730), 371.
23. Newton, *Optice* (1706), qu. 23, 338–9, especially 338: '*Vis repellens*', 'consequi videtur'; compare id., *Opticks* (1730), 395: 'repulsive Virtue'. Newton, *Principia* (1687), vol. 1, 429–31.
24. Newton, *Optice* (1706), qu. 23, 344–5. Id., *Principles*, vol. 2, General Scholium, 546–7.
25. Id., *Principia* (1687), vol. 1, 339–44, especially 343: 'quasi magis attracti' and 344: 'de natura radiorum (utrum sint corpora necne) nihil omnino disputans'. Compare Newton, *Principles*, vol. 1, 230 and 230–1.
26. Id., *Optice* (1706), qu. 21, 316–17.
27. As early as 1675–6, a manuscript of Newton's on an ether, in explanation of optical phenomena, was read to the Royal Society. This text was not published until 1757. Birch, *History*, vol. 3, 247–305.
28. Newton, *Opticks* (1730), qu. 29, 374; qu. 18–20, 348–50. The fourth English edition (1730) did not deviate on any essential points from the second English edition of 1717; Koyré, 'Études'.
29. Newton, *Opticks* (1730), qu. 21–2, 350–3; quotations 352. Newton indicates that there are actually extremely rarified media: air at a great height and electrical and magnetic effluvia; ibid., 353.
30. See Section 3c below. For the status of forces conceived by Newton, 's Gravesande and van Musschenbroek, see de Pater, *Musschenbroek*, 88–95.
31. Cantor, *Optics*, 69–70: six of the sixty-seven 'projectile theorists' utilised a dynamic ether. Cantor provides no comparable data on the explanation given by

the twenty-two 'fluid theorists' (ibid., 91–113, 208–9). See also note 3 to Chapter 5.

32. See Section 3b below.
33. De Mairan, *Dissertation*. De Mairan's ideas on luminescence are discussed by Harvey, *History*, 151–4.
34. De Mairan, *Dissertation*, 2–3.
35. Ibid., 10–11.
36. Ibid., 5.
37. Malebranche, *Oeuvres*, vol. 3, 255–8. See also Section 2b below.
38. De Mairan, *Dissertation*, 6 (in the original the whole passage is given in italics).
39. Ibid., 6–9.
40. Ibid., 6, 9–10.
41. Van Musschenbroek, *Epitome*, 246.
42. This is not to say that 's Gravesande and van Musschenbroek did in fact base their accounts on de Mairan.
43. De Mairan, *Dissertation*, 11–14, 19, quotations 19, 11 (the last quotation is given in italics in the original).
44. Ibid., 15–16.
45. Ibid., 16–19.
46. Ibid., 20–2, 24, quotation 22 (in the original, 'principe actif' is given in italics).
47. Ibid., 34, 36, 45.
48. Ibid., 43–5.
49. Newton's second English edition of the *Opticks*, which includes the ether hypothesis, appeared in 1717. See Section 1a, above.
50. De Mairan, 'Suite', 369–70, 372–5, especially 373.
51. Ibid., 375–6.
52. De Mairan, *Dissertation*, 50–52.
53. Ibid., 46–50, quotation 48.
54. Ibid., 50: 'Je dois avertir aussi que cette experience [the formation of a colour spectrum by a prism] ne réussit qu'à la Lumiere du Soleil, parce que, *comme je l'ai souvent éprouvé*, les rayons de tout autre Lumiere sont de beaucoup trop foibles, lorsq'on éloigne le prisme du corps lumineux; ou trop divergens lorsqu'on l'en approche' (emphasis added: CH). Guerlac, *Newton*, 115–16, supports the thesis that de Mairan performed these tests, curiously enough, exclusively with reference to the foreword to the second French translation of the *Opticks*, 1722. Kleinbaum, *Mairan*, 151, cites – in another context – a letter of 1718 in which de Mairan talks about his experiments. Both of them appear to have overlooked the printed passage of 1717.
55. 's Gravesande, *Elementa* (1720–1). The survey article in Zedler's encyclopaedia refers to Newton's *Opticks* and 's Gravesande's *Elementa*; *Grosses Lexicon*, vol. 17 (1738), 828.
56. 's Gravesande, *Elementa*, vol. 2 (1721), 1–2.
57. Ibid., 11: 'Corpus *vocatur* lucidum, *quod lumen emittit*, id est, ignem per lineas rectas agitat' and ibid., 10: '*Quando ignis per lineas rectas oculos nostros intrat . . . ideam luminis excitat*' (emphasis in the original).

58. Ibid., 17–19.
59. Ibid., 59–64, especially 63: '[*L*]*umen repercuti ad certam distantiam à corporibus*, eodem modo ac vis refringens ad certam à corpore distantiam agit' (emphasis in the original).
60. Ibid., 63, 98–103.
61. 's Gravesande was probably not unfamiliar with the ether hypothesis, which Newton had published in 1717. From a letter dated 24 June 1718 from 's Gravesande to Newton it appears that he possessed the second English edition of the *Opticks* (probably a copy of the second impression of 1718); Hall, 'Further', 26.
62. 's Gravesande, *Elementa* (1742), vol. 2, 701: '*Si per foramen Lumen transeat, directionem servat, & non ad latera dispergitur*, ut de Undis dictum' (emphasis in the original).
63. Ibid., 702.
64. Ibid., 702–3. In 1705 George Cheyne actually described an experiment with a very small hole in a vertical plate, claiming that it showed the extreme minuteness of the light particles; Cantor, *Optics*, 32.
65. Ibid., 703–14. In 1676 Römer determined the speed of light on the basis of the eclipses of Io, one of Jupiter's moons. In 1729 Bradley obtained a value for the velocity of light by analysis of the astronomical aberration he had discovered.
66. 's Gravesande knew Huygens' *Traité* well. In 1728 he had published a Latin translation of the *Traité*. However, he gives no judgement here on Huygens' theory of light; Huygens, *Opera*, vol. 1, preface 4, text 1–92.
67. 's Gravesande, *Elementa* (1742), vol. 2, 655, 657, 725, especially 657: '*Lumen ex Corporibus per Lineas rectas emittitur*' (emphasis in the original).
68. Ibid., 725–6, 733, 839.
69. Ibid., 733; vol. 1, 24.
70. See de Pater, *Musschenbroek*, 29–32.
71. Van Musschenbroek, *Elementa* (1741); id., *Grundlehren*.
72. Id., *Epitome*; id., *Elementa* (1734).
73. Id., *Elementa* (1741), 309, 325–6.
74. Ibid., 354–5, 358–9.
75. Ibid., 359–60.
76. See Chapter 5, Section 4.
77. Van Musschenbroek, *Elementa* (1741), 327, 338–9, especially 338: 'effectus . . . ultra fidem magni'.
78. Ibid., 351. In 1739 he accepted the possibility that there might be other kinds of fire in which the 'electric fire' would constitute a separate kind; van Musschenbroek, *Beginsels*, 506.
79. Van Musschenbroek, *Elementa* (1741), 362, 370–1.
80. Ibid., 183–5.
81. Ibid., 387–8. In 1727 van Musschenbroek mentioned a variable speed as a third possibility; *Epitome*, 267–9.
82. Van Musschenbroek, *Elementa* (1741), 391–2.
83. Id., *Elementa* (1734), 342–3. Van Musschenbroek probably derived the idea for Figure 5 from a similar depiction in Desaguliers, 'Optical experiments', opposite p.575.

84. See Section 1a, above.
85. Van Musschenbroek, *Elementa* (1741), 355–8. There are no references to the literature in 1741. In 1727 there was no passage on the sun and the horizon; for further information readers were referred to Newton's *Principia* and 'Dortous de Mairan, Sur la Lumiere' (*Epitome*, 246). See also Section 1b, above.
86. Van Musschenbroek, *Elementa* (1741), 365–7, 388–9.
87. Ibid., 354, 358–9.
88. Van Musschenbroek, *Beginsels*, 518–21, quotation 519; id., *Elementa* (1741), 359–61.
89. Descartes, *Principia*, 108–16. Dijksterhuis, *Mechanisering*, 451; Shapiro, 'Propagation', 243–52.
90. Descartes, *Principia*, 116–37. Shapiro, 'Propagation', 252–4.
91. He did explain the light of comets' tails by a new kind of refraction that, however, was based upon principles completely different from those obtaining for normal refraction; Descartes, *Principia*, 185–92. Shapiro, 'Propagation', 292–4.
92. Descartes, *Dioptrique*, 88–9. The comparison to the motion of a ball can be found in writers as early as Ptolemy, Alhazen, and others; Sabra, *Theories*, 93–9.
93. Descartes to Mersenne, 5 October 1637; Descartes, *Oeuvres*, vol. 1, 451.
94. Sabra, *Theories*, 300–8, discusses the similarity between the deductions made by Descartes and Newton.
95. Descartes, *Dioptrique*, 89–92 and *Meteores*, 331–5, especially 333.
96. See, for example, Malebranche (Section 2b, below).
97. Shapiro, 'Kinematic optics', 137, 207–58. Here I use 'pulse front' in place of 'wave front' in order to avoid losing sight of the fact that Huygens' pulses are nonperiodic.
98. Huygens, *Traité*, 10–16, especially 15–16.
99. Ibid., 17–18, especially 18: 'infinement'.
100. Ibid., 19–20.
101. Shapiro, 'Kinematic optics', 225–7, shows with great precision that Huygens' principle cannot explain rectilinear propagation, even if only Huygens' own concepts and criteria are used. Shapiro's proof is criticised by de Lang, "Originator", 21, 24–7.
102. Huygens, *Traité*, 48–9, 58–62; for the designation *experimentum crucis*, see page 54. Shapiro, 'Kinematic optics', 236–44.
103. Shapiro, 'Kinematic optics', 245–6, posits that Philippe de la Hire, Denis Papin, and Leibniz were the only ones in the seventeenth and eighteenth centuries to accept Huygens' principle.
104. Ibid., 253–5, on Antoine Parent. For Johann II Bernoulli's criticism see Section 2b, below; for Euler's, see Chapter 4, Section 3a. Shapiro places too much stress on the relationship between the rejection or ignoring of Huygens' explanation of double refraction and the rejection or ignoring of his theory (ibid., 245, 252); Huygens' problems with the explanation of rectilinear propagation were primarily responsible for the latter.
105. Shapiro, 'Kinematic optics', 244. See Section 2b, below, and Chapter 4, Section 1.

106. Huygens, *Traité*, 3. Huygens to Leibniz, 29 May 1694; Huygens, *Oeuvres*, vol. 10, 612–13, quotation 613.
107. Huygens, *Discours*, 161–3.
108. Ibid., 163–5.
109. Newton, *Principia* (1687), vol. 1, 514: 'Hoc experimur in sonis, qui vel domo interposita audiuntur, vel in cubiculum per fenestram admissi sese in omnes cubiculi partes dilatant, inque angulis omnibus audiuntur, non reflexi a parietibus oppositis sed a fenestra directe propagati.' Compare id., *Principles*, vol. 1, 370.
110. Huygens, *Discours*, 164.
111. Id., *Oeuvres*, vol. 19, 372; the experiment described dates from 1669.
112. Newton, *Principia*, vol. 1, 514: 'monte interposito', 'quantum ex sensu judicare licet'; compare id., *Principles*, vol. 1, 370; id., *Optice* (1706), qu. 20, 307. See also Newton to Leibniz, 16 October 1693: 'motus sonorum forte magis rectilineus est'; Newton, *Correspondence*, vol. 3, 285–6.
113. Robinet, *Malebranche*, 261–323; Duhem, 'L'optique', 74–91.
114. Malebranche, *Oeuvres*, vol. 2, 328 and 335 (*Recherche*, first edition 1675). Id., *Oeuvres*, vol. 12, 279–80 (*Entretiens*, first edition 1688).
115. Before this lecture was published in 1702 in the memoirs of the Paris Academy, the fifth edition of the *Recherche* appeared in 1700, with a somewhat different version of the lecture, as 'Eclaircissement sur la lumiere & les couleurs, & sur la generation du feu'; Malebranche, *Oeuvres*, vol. 3, 253–69 (variorum edition).
116. Malebranche, *Oeuvres*, vol. 3, 261. Robinet, *Malebranche*, 283–4.
117. Malebranche, *Oeuvres*, vol. 3, 263–4 (1700).
118. Ibid., 257, 260, 266 (1700).
119. Ibid., 258–60 (1700).
120. Robinet, *Malebranche*, 266–76. There are no indications that Malebranche was guided by Newton's suggestion to Hooke.
121. Malebranche, *Oeuvres*, vol. 3, 253–305, 306–48 (*Recherche*, sixth edition 1712, 'XVI. Éclaircissement' and 'Dernier éclaircissement'). For the influence of Newton's *Optice*, see Robinet, *Malebranche*, 299–323.
122. Malebranche to P. Berrand, 1707; Malebranche, *Oeuvres*, vol. 19, 771–2.
123. Malebranche also shows his admiration in print. He talks of the 'expériences que M. Newton, ce sçavant Géometre & si renommé en Angleterre & par tout, a faites avec une exactitude telle que je ne puis douter de la vérité'; ibid., vol. 3, 302 (*Recherche*, sixth ed., 1712).
124. Ibid., 258, 300–2 (1712), especially 302: 'le rayon rouge, qui a le plus de force, puisqu'il souffre moins de réfraction que les autres rayons, n'est pas repoussé si promptement, ou recommence ses vibrations moins souvent que ceux qui le suivent; & que le violet qui est le dernier & le plus foible, est celui de tous dont les vibrations sont les plus petites & les plus promptes, ou recommencent plus souvent.'
125. Ibid., 298–9 (1712).
126. Ibid., 299–301 (1712).
127. Malebranche says about his optical ideas: '[J]e croi devoir avertir qu'on ne doit regarder que comme des conjectures ou des vûës generales insuffisam-

ment prouvées, ce que je viens de dire . . . , pour rendre raison des principales expériences que M. Newton . . . a faites'; ibid., 302 (1712).

128. Bernoulli, 'Recherches'.
129. Brunet, *L'introduction*, 309–18; Latchford, *Young*, 79–83; Whittaker, *History*, vol. 1, 95–6. (On p.96 Whittaker regrets that Bernoulli narrowly missed making a great discovery, namely, the transversality of light vibrations. Bernoulli, however, was comparing only the transversal vibrations of a cord with the longitudinal vibrations of light; for him and his contemporaries the idea of transversal light vibrations was absurd, as it still was eighty years later.)
130. Cannon and Dostrovsky, *Evolution*, 77–82; Truesdell, "Editor's introduction", xxx–xxxii.
131. Bernoulli, 'Recherches', 8–9, 41–2.
132. Ibid., 10.
133. Ibid., 10–12, 25–6, quotation 10.
134. Ibid., 10–15.
135. Ibid., 15, 55.
136. De Mairan presented this idea in 1719 to the Academy; Fontenelle, 'Observations', 11; de Mairan, 'Discours', 3.
137. Bernoulli, 'Recherches', 13.
138. Ibid., 25.
139. Ibid., 29–30, 35.
140. For the mathematical elaboration of Bernoulli's theory, see Cannon and Dostrovsky, *Evolution*, 47–52, 77–82.
141. Bernoulli, 'Recherches', 43–4.
142. Bernoulli, 'Disquisitio'.
143. Bernoulli, 'Recherches', 44–7.
144. Ibid., 47–50, quotation 48.
145. Ibid., 52–5.
146. Ibid., 56–62.
147. Ibid., 62–6, quotation 64.
148. Another example concerns the explanation of the homogeneity of a ray of light, which, according to Bernoulli, is not satisfactorily explained by Newton's emission conception; ibid., 55–6.
149. Ibid., 7, 42, quotation 7.
150. See Chapter 1, Section 1.
151. Shapiro, 'Kinematic optics', 153–5, 172, 179.
152. Béguelin, 'Recherches' (1774), 154–5.
153. De Mairan, 'Recherches', 49–50. He recalls this thesis in 'Suite', 366, and 'Discours', 24.
154. Id., 'Suite', 366.
155. Id., 'Troisième partie', 4–5, 32.
156. Hall, 'France', 242; id., 'Summary', 309–10; Elkana, 'Newtonianism'. Heilbron, *Electricity*, 1–2; id., 'Experimental', 361; Home, 'Magnet', 263. See also Home, 'Straitjacket', 235–44; Hakfoort, 'Etiketten'; id., 'Wolff', 36–8.
157. Winkler, 'Institutiones', 334–5. See Chapter 5, Sections 1a and 1b.

4. Euler's 'Nova theoria' (1746)

1. The publication of Euler's collected works began in 1911 and is planned to reach around eighty-five volumes, of which more than seventy have appeared. The works are divided into four series: mathematics; mechanics and astronomy; physics and various subjects; letters and manuscripts. Eneström, *Verzeichnis*, gives an inventory of Euler's printed works. An introduction to Euler's life and work is provided by Burckhardt et al., eds., *Euler*, and the literature given there; Thiele, *Euler*; Juškevič, 'Euler'.
2. See Fellmann, 'Stellung' and the literature given there. After the first (Dutch) edition of the present work, Home, 'Leonhard', appeared, discussing aspects of Euler's wave theory of light.
3. Euler, *Opera*, vols. III.6–III.9, III.3–III.4, III.5, respectively.
4. Id., *Lettres*. Id., *Opera*, vol. IVa.1 provides summaries of all the known letters to and from Euler. The 'Nova theoria', the *Opuscula* (in which the article was originally incorporated), or Euler's wave theory in general are discussed in the exchange of letters with twenty-four of his correspondents.
5. For the study of Euler's physical optics Speiser's summarising "Einleitung" in Euler, *Opera*, vol. III.5, is extremely useful. See also Speiser, 'L'oeuvre', and id., 'Schriften'.
6. Speiser, 'Einleitung', liii, is an important exception. See now also Home, 'Leonhard' (compare note 2, above).
7. For Huygens, see Chapter 3, Section 2a. For Euler, see Section 3a, below.
8. See Section 3a, below.
9. Shapiro, 'Huygens' theory', 206–7.
10. Euler's manuscript 'Catalogus librorum meorum', which can be dated to 1747, or shortly after, contains 509 numbers. Only an edition of Newton's *Lectiones opticae* and the *Principia* are cited. There is no mention of works by Descartes, Huygens, Malebranche, de Mairan, or Johann II Bernoulli. Yet this is not to say that Euler had none of these authors' works in his possession. It is clear, for instance, from an exchange of letters with Johann I Bernoulli (the father of Johann II), that Euler acquired a copy of Johann II's 'Recherches' shortly after it appeared, and provided a commentary on this in a letter to Johann I (Johann I Bernoulli to Euler, 2 April 1737, and Euler's answer of 27 August 1737; Eneström, 'Briefwechsel', 251, 256). For Malebranche and Huygens one can say that Euler knew these authors at least indirectly, through the work of Johann II. It is clear that he was well aware of Descartes' and de Mairan's work, from passages in the 'Nova theoria' (pp. 9, 21), where he gives a critique of their ideas.
11. See Sections 3d and 3e, below.
12. Euler, 'Pensées'. This manuscript is the text of a lecture given on 6 February 1744 to the Société Littéraire, which functioned in the period 1743–5 as the Berlin Academy of Science. According to the minutes, on that day Euler read a 'dissertation sur la lumiere et les couleurs' (archives, Akademie der Wissenschaften, Berlin, no. I-IV-10, p. 20). A summary of the lecture, compiled by J. H. S. Formey, appeared in the memoirs of the Berlin Academy (Euler, ed. Formey, 'Sur la lumiere'; for the attribution of the summary to Formey, see

Winter, ed., *Registres*, 36–7 (I am obliged to B. Bosshart for this reference). The text of the manuscript is noticeably longer than the published summary: It contains more examples and mathematical deductions, the latter being entirely absent from the summary. There are also significant differences in the choice of words, which, generally speaking, do not affect the tenor of the piece.

13. Euler, 'Disquisitio', 289, concerning the law of dispersion. In two previous cases astronomical aberration had been the subject: Euler, 'Explicatio' and 'Mémoire'. See Speiser's comments, 'Einleitung', xv–xx, xlvii–il.
14. Daniel Bernoulli to Euler, 29 April 1747; Euler, *Opera*, vol. IVa.1, 37.
15. Euler, 'Nova theoria', 1: 'Omnem sensationem fieri per contactum, quo in nostro corpore mutatio quaedam producatur, tam ratio quam experientia ita dilucide docet, ut nullum amplius dubium superesse possit.' There is a French translation of the 'Nova theoria', but it has to be used with caution, because of omissions, additions, and outright mistakes; Euler, *Oeuvres*, vol. 2, 362–417.
16. Euler, 'Nova theoria', 1: 'vel . . . ab his corporibus effluvia emanant, atque sensuum nostrorum organa feriunt, vel in circumiacentibus corporibus eiusmodi motionem excitant, quae ad nostros sensus usque per omnia corpora intermedia propagetur.'
17. Ibid., 2: 'plerique alii Philosophi'.
18. Ibid.: 'Quamvis . . . summum cuiusque theoriae firmamentum in perfecta omnium phaenomenorum explicatione . . . sit positum, tamen ne dubitatione in ipso limine concepta animum lectoris diutius ambiguum teneat, operam dabo, ut alteram sententiam . . . non solum altera magis probabilem, sed etiam veritati prorsus consentaneam ostendam.'
19. Ibid.: 'ob maiores distantias'; 'iis, qui contrariae opinio sunt addicti, vix attentione dignum videri solet; quamobrem ipsa fundamenta, quibus hi suam sententiam confirmare conantur, potissimum erunt examinanda ac labefactanda.'
20. Deleted: 'la'; i.e., the earlier version is 'de la ressemblance'. For this quotation but also in general, the alterations do not affect the drift of the text. They are intended as improvements of Euler's French. The corrections are by Euler's colleague P. Naudé (Euler to C. Goldbach, 4 July 1744; Juškevič and Winter, eds., *Euler und Goldbach*, 199; I am indebted to B. Bosshart for this reference).
21. Deleted: 'quel'.
22. Difficult to read: 's'assurer'.
23. Difficult to read: 'assez'.
24. Deleted: 'vient'; replaced by 'est parti'.
25. The sentence has been changed by deletions and additions. The original version reads: 'Telle ressemblance si étroite ne nous laisse pas douter'.
26. Added: 'ainsi'.
27. Euler, 'Pensées', section 1.
28. Id., ed. Formey, 'Sur la lumiere', 17.
29. Id., 'Nova theoria', 3: 'coelos omnis resistentiae expertes atque adeo vacuos'.
30. Ibid., 3–4, quotation 3: 'omnem imaginationis vim superat'.
31. See page 32.
32. The argument about the great velocity had already been used at an earlier date, for instance, by Huygens and Wolff (see Chapter 3, Section 2a, and Chapter 5,

Section 1a, respectively). Euler was to mention it as a separate argument later on; Id., *Lettres*, vol. 1 (1768) (*Opera*, vol. III.11, 47).

33. See pages 29–31 and 55.
34. Ibid., 4–5.
35. Ibid., 5–6.
36. Ibid., 18–19. For Newton's remark, see page 30, where I also noted that Newton gave a theoretical proof in the *Principia* of 1687. In 1766 Euler showed why, for the one-dimensional case, a pulse is only propagated in one direction; Euler, 'Propagation du son', 444–8.
37. See Chapter 5, Section 3a.
38. Id., 'Nova theoria', 6–7, quotation 7: 'Etsi non ignoro patronos emanationis radiorum in magnitudine huius stupendi numeri nihil absurdi invenire, tamen non dubito, quin ob hoc ipsum ista opinio non parum de sua probabilitate apud aequos iudices sit amissura.'
39. See pages 40 and 48, respectively. Segner's treatise of 1740, which will be discussed in the context of Euler's fifth argument, is also relevant. For the theme of the subtlety of light see: Schagrin, 'Experiments', 222–9, and Cantor, *Optics*, 33, 52–9.
40. Euler, 'Nova theoria', 7.
41. Newton, *Opticks* (1730), qu. 16, 347.
42. Segner, *De raritate*, 4–8.
43. Melvill, 'Observations', 14–17; Canton, 'Method', 343–4. Schagrin, 'Experiments', 226–7; Cantor, *Optics*, 54–5. Rosenberger, *Geschichte*, vol. 2, 345. Segner's treatise is also mentioned in Klügel's German translation of Priestley's *History*: Priestley, *Geschichte*, vol. 2, 280.
44. Euler, 'Nova theoria', 7; id., 'Pensées', section 11.
45. Newton, *Opticks* (1730), 267–9. For this 'nutshell' conception of matter, see Thackray, 'Matter', and id., *Atoms*, 53–67.
46. Bošković, *Theoria*, 220–5; id., *Theory*, 168–72.
47. Priestley, *History*, vol. 1, 390–4.
48. Euler, 'Nova theoria', 7–8.
49. Ibid., 9: 'formatione ac propagatione pulsuum'; 19: 'De pulsuum successione atque radiis lucis'; 25: 'De reflexione et refractione radiorum'; 34: 'corporibus lucentibus, reflectentibus, refringentibus et opacis'.
50. Id., ed. Formey, "Sur la lumiere", 19 (emphasis in the original).
51. Euler, 'Nova theoria', 9. Two years earlier Euler advanced the objection to the globule models that they did insufficient justice to the uniformity of the ether; id., 'Pensées', section 7.
52. Id., 'Nova theoria', 8.
53. Id., *Anleitung*; for the dating see Truesdell, 'Rational fluid mechanics', c. For the ether conception in Euler's 'De sono' and 'Pensées', see Home, 'Leonhard', 529–33.
54. Euler, *Anleitung*, 103–4, 107–10, 113–14, 118. Parts of Euler's ideas on the ether found their way into the same collection in which 'Nova theoria' appeared; see especially 'Moindres parties'. Cantor, *Optics*, 118, writes – wrongly – that Euler's ether is 'particulate'. True, Euler does talk in the 'Nova theoria' about 'particula', or parts of the ether, but by this he means no fixed

or smallest parts. The 'parts' are mathematical 'regions' of ether, as when we talk about 'parts' of space.

55. Euler, 'Nova theoria', 12; Newton, *Principia*, vol. 1, 521–8. On Newton's treatment, see Truesdell, 'Introduction', xxi–xxiv; Cannon and Dostrovsky, *Evolution*, 9–14.
56. Euler, 'Nova theoria', 9–12.
57. Ibid., 12–13. In the original a/T and c stand for v and L, respectively. In the following the notation is also tacitly adapted; however, conceptual changes with regard to Euler's treatment are explicitly mentioned.
58. See note 54, above.
59. Ibid., 12–15; Newton, *Principia*, vol. 1, 524. For Euler's treatment, compare Truesdell, 'Introduction', xxxii–xxxiii; Cantor, *Optics*, 119–20. In the original, formula (11) has $\frac{1}{2}k$ rather than k. Euler's factor $\frac{1}{2}$ has no real meaning and is based upon his special choice of units; Truesdell, 'Rational fluid mechanics', xliii–xliv.
60. Euler, 'Nova theoria', 15.
61. At the beginning of the nineteenth century Laplace advanced the suggestion that an increase in temperature resulting from adiabatic compression could explain the difference between theory and experiment. Fox, *Caloric*, 81–6; Kuhn, 'Caloric', 136–8.
62. Euler, 'Nova theoria', 15–16.
63. Euler uses the values (rounded off by me): speed of sound = 338 m/s, speed of light = 2.11×10^8 m/s. The value used for the speed of light, low even for that time, resulted from Euler's use of an exceptionally high value for the solar parallax; Boyer, 'Early', 36.
64. Euler, 'Nova theoria', 16–17; id., 'De relaxatione'.
65. Van Musschenbroek, 'Introductio ad cohaerentiam', especially 494–507. Id., *Elementa* (1734), 168–9, and *Elementa* (1741), 222–3.
66. $F_{weights}$ is inversely proportional to the surface, according to van Musschenbroek, *Elementa* (1741), 222.
67. Id., *Introductio ad philosophiam*, vol. 1, 416–17. Euler, 'Nova theoria', 17–18.
68. This kind of consideration seems to have played a role for Young when he dropped the idea, which he once cherished, that the light ether also causes the cohesion of bodies; Cantor, 'Changing', 56–8.
69. Euler, 'Nova theoria', 19–20.
70. See Section d, below.
71. See note 10, above.
72. Euler calls this distance c. In view of the later established custom of designating the speed of light by this symbol, I have replaced it with the symbol d.
73. Euler, 'Nova theoria', 22–3.
74. Ibid., 23.
75. Ibid., 24–5, 26, quotation 24.
76. Ibid., 23–4.
77. Ibid., 24; 31–2, see Section d, below.
78. See page 63. Shapiro, 'Kinematic optics', 212.
79. Euler, 'Nova theoria', 25–6.

80. Newton, *Opticks* (1730), 262–6.
81. See Section e, below.
82. Shapiro, 'Kinematic optics', 255.
83. Euler, 'Nova theoria', 26–7.
84. Ibid., 27–8.
85. For the history of this problem in Britain, see Cantor, *Optics*, 64–9.
86. Euler, 'Nova theoria', 29.
87. Ibid., 29–30.
88. Ibid., 30. For Malebranche, see Chapter 3, Section 2b.
89. Euler, 'Coniectura', 128, and 'Essai', 171 (violet has the greatest frequency); id., 'Réfrangibilité', 221 (red has the greatest frequency); id., *Lettres*, vol. 2 (1768), *Opera* edition, vol. III.12, 7 (it is not known which colour has the greatest frequency).
90. Gehler, *Wörterbuch*, vol. 2 (1789), 150, 899; Heinrich, 'Preisfrage', 262–4.
91. Euler, 'Nova theoria', 34: 'sine dubio'.
92. Ibid., 31–2.
93. Ibid., 32–3.
94. Ibid., 33.
95. Ibid., 34–6.
96. Ibid., 36: 'non multum a celeritate in aethere discrepantem'.
97. Ibid., 37.
98. Ibid., 39; Newton, *Opticks* (1730), 179–85.
99. Newton, *Opticks* (1730), 251–5.
100. Euler, 'Nova theoria', 39–40. Two years earlier Euler noted, as a third possibility, the explanation given by the 'Cartesians'. He described this as the idea that colours are a mixture of light and shadow. This conception became untenable, in Euler's view, after 'la diversité des rayons' had been demonstrated; Euler, 'Pensées', section 18.
101. Euler, 'Nova theoria', 40–1.
102. Ibid., 42–3.
103. Euler, 'Pensées', section 19.
104. Id., 'Nova theoria', 44: 'Conveniunt enim mirifice omnia reliqua phaenomena, quae hae corporum alterationes suppeditant'.
105. Ibid., 44–5.
106. Ibid., 41: 'Ponamus radium, qui colorem rubrum repraesentat, uno minuto secundo ad oculum afferre f pulsus; et quemadmodum in musica soni, quorum vibrationes eodem tempore editae rationem tenent duplam vel quadruplam vel octuplam vel etc. pro similibus habentur, ita quoque radii simplices, qui uno minuto secundo vel 2f, vel 4f, vel 8f etc. vel etiam $^1/_2$f, vel $^1/_4$f, vel ($^1/_8$)f etc. vibrationes continent, omnes rubri censebuntur.' N.B. In the original the symbol α is used for frequency.
107. Id., 'Pensées', section 15.
108. Id., 'Nova theoria', 42–3, quotation 42; id., 'Pensées', section 15.
109. Young, 'Theory', 47; id., 'Experiments', 15–16.
110. Cantor, *Optics*, 130.
111. In a treatise on sound published in 1766 Euler explicitly abandoned the idea that underlies the explanation of dispersion; he no longer believed that an in-

crease in speed occurred because one pulse was followed by another; Euler, 'Propagation du son', 450.

112. Compare Buchwald, *Rise* 3–8.
113. See Chapter 5, Section 3a.

5. The debate in Germany on the nature of light, 1740–95

1. Ronchi, *Histoire*, 218; Wolf, *History*, 163.
2. Weinmann, *Natur*, 114, mentions (besides Euler) J. C. P. Erxleben; Klemm, *Geschichte*, 23–24, mentions M. Knorr, C. Wolff, and, as adherents of Euler's wave theory, Erxleben, G. S. Klügel and W. J. G. Karsten (the classification of Klügel is incorrect; see pages 142–3).
3. Cantor, *Optics*; the figures are calculated with the aid of the data in the appendix on pp. 206–11. I have added the forty-five supporters of a 'projectile theory' to the eighteen adherents of a 'fluid theory'. Although Cantor rightly stresses the differences between the two conceptions, they can both be regarded as belonging to the emission tradition (ibid., 13, 91). The existence of the relatively strong support for the 'fluid theories', which diverge markedly from the 'projectile theories' of the 'Newtonians', supports the thesis that the emission tradition, even in Britain and Ireland, did not possess a monolithic appearance.

   Cantor writes that the adherents of a 'fluid theory' had no great interest in the usual optical problems (reflection, refraction, and so on). They were primarily interested in general cosmological and theological questions. To a large extent they therefore placed themselves outside the debate on the nature of light, at any rate the debate that was carried on by the supporters of a 'projectile theory' and others. For this reason we might regard the 'fluid theoreticians' as a group that had in fact elected for neither the emission nor the medium tradition. If we would do this, the eighteenth-century support for emission and medium traditions would be 88 and 12 per cent, respectively. This alternative method of calculation produces no really different picture. In my view, Cantor goes too far when he combines the support for the 'fluid theories' and that for the medium conceptions, arriving at a figure of 35 per cent for nonsupporters of the 'projectile theory'. He regards this as an argument for the idea that the emission tradition in the eighteenth century was not dominant (ibid., 200). N.B. There is an element of uncertainty in all figures given: Cantor provides no data on the people who made no choice in favour of one of the available conceptions, although the debate on the nature of light was relevant to them, or who even supported a completely different conception.
4. Gehler, *Wörterbuch*, vol. 2 (1789), 897, 902. For Gehler's mild preference for an emission theory, see Section 6, below.
5. Fischer, *Geschichte*, vol. 4, 448.
6. In most cases the name and the year, given in the table, can identify the source in the bibliography. Supplementary information is needed in the following cases.

   Kaschube 1718 = id., *Elementa*. *Grosses Lexicon* 1738 = article 'Licht', ibid., vol. 17 (1738), 825–8. Klügel 1776 = Priestley, ed. Klügel, *Geschichte*. Werner 1788 = id., *Entwurf*, and 'Versuch'. Gehler 1789 = id., *Wörterbuch*, vol. 2 (1789), article 'Licht', 882–904. Gren 1790 = Karsten, ed. Gren, *Naturlehre*. Kries 1792 = Euler, ed. Kries, *Briefe*, vol. 1 (1792), 201–46. Lichtenberg 1791 = Erxleben, ed. Lichtenberg, *Anfangsgründe* (1791). Gehler 1795 = id., *Wörterbuch*, vol. 5 (1795), supplement to the article 'Licht', 546–56.

In column 2 the only sources included are those in which the discussion on light was found relevant but in which no choice was made. Therefore, in most cases sources of the following kind are not included: textbooks on applied mathematics with chapters on geometrical optics, treatises on technical aspects of optics, and writings examining the empirical aspects of the theory of colours. Examples of these types of sources are, respectively, Kästner, *Anfangsgründe*; Zeiher, *Abhandlung*; Goethe, *Beiträge*.

7. Exceptions: Kant, *De igne*; Lichtenberg, 'Schreiben'.
8. Scheibel, *Einleitung*, vol. 9 (1777); Murhard, *Litteratur*, vol. 5 (1805); Ersch, *Handbuch*, vol. III.2 (1828); Goethe, *Schriften*, vol. I.6, 344–51; vol. II.6, 91–110, 536–40.
9. On the basis of a radical change in the argumentation the following were also included: Karsten, ed. Gren, *Naturlehre*, and Erxleben, ed. Lichtenberg, *Anfangsgründe* (1794). See Section 6, below.
10. Twice: Gehler, Gren, Lichtenberg, Segner. Three times: Eberhard.
11. An important exception is Segner, *De raritate*.
12. *Grosses Lexicon*, vol. 9 (1735), 223–40.
13. Ibid., vol. 17 (1738), 823–8.
14. Wolff, *Würckungen der Natur* (1723). Hamberger, *Elementa*. For the importance of Wolff and Hamberger, see Stichweh, *Entstehung*, 99; Heilbron, *Electricity*, 141.
15. Kaschube, *Elementa*; Winkler, *Institutiones*. Segner, *Einleitung* (1746); see also Segner, *De raritate*. In Kaschube, *Cursus* (2nd edition 1718), 232–3, the author finds it plausible that the 'ether particles' differ in size and that particle size determines the colour of the light. Since he also refers the reader to Newton's *Opticks* for further information, the impression might be given that Kaschube supports an emission conception. In view of the medium conception in the *Elementa* (also dating from 1718), I assume that Kaschube wants to give a mechanistic explanation of colours in the *Cursus* that fits in with a medium conception. The idea underlying this explanation is, however vaguely formulated and undeveloped, to a certain extent related to Johann II Bernoulli's 'fibre theory' (see Chapter 3, Section 2b).
16. Wolff, *Würckungen der Natur*, 164–7, 176–82, especially 179–80.
17. Ibid., 178–9.
18. Ibid., 180–7.
19. Ibid., 187–8.
20. For refraction, for example, see Wolff, *Anfangs-Gründe*, vol. 3, 6–7, 67–9. In his 'Algebra' Wolff provides a deduction from the refraction law based on the principle of shortest time, in which he implicitly assumes that the speed is less in glass than it is in air; ibid., vol. 4, 368–9.
21. Kaschube, *Elementa*, 100.
22. Winkler also refers to Huygens without discussing Huygens' principle; Winkler, *Institutiones*, 301.
23. Wolff, *Würckungen der Natur*, 188–90.

24. Winkler, *Institutiones*, 334–5; id., *Anfangsgründe* (1754), 45, 252.
25. Hamberger, *Elementa* (1727), for example, 287–9 ('s Gravesande); 413 (van Musschenbroek); 199, 200, 205 (Newton).
26. Ibid., 125, 188–9.
27. No argumentation: Hollmann, *Primae*; only rectilinear propagation: Segner, *Einleitung* (1746), 247–8.
28. Hamberger, *Elementa* (1727), 191–2.
29. Ibid., 211–12; 125: '[Ignis] particulas habet sphaericas, . . . quae figura etiam ex legibus reflexionis, . . . quas servant, (§. 468.) demonstrari potest'. (In section 468 the law of the equal angles of incidence and of reflection is explained.)
30. Ibid., 202–5.
31. Mangold, *Systema*, regards the problem of the nature of light as a choice between four different theories. See Section 2, below.
32. Here two other private notes by Lichtenberg can be mentioned. The first (from the period 1789–93) contains a suggestion on the direction in which an entirely new theory of light may be sought: 'Sollten wir nicht Affinitäten gegen andere Körper fühlen können? und sich nicht daraus eine dritte Theorie des Lichts herleiten lassen. Was ist Affinität bei einem empfindlichen Geschöpf? Ist nicht unsere ganze Relation unserer Anschauungen nur verändert das, was in der Chymie Affinität heisst? N.B. Licht, Schall, ist *gefühlte* Affinität' (Lichtenberg, *Schriften und Briefe*, vol. 2, 313; emphasis in the original). The second note (from the period 1793–6) appears to be a half-serious, half-ironic plea for a theory that gives rise to a synthesis between emission and medium traditions: 'Das "medium tenuere beati" ist so abgebraucht, dass man nun allmählich anfangen kann es wieder für brauchbar zu halten. Wie wäre es, wenn man am besten damit auskäme, beide Theorien des Lichts, die Newtonische und die Eulerische, zu vereinigen? Ueberhaupt ist das medium tenuere beati eine goldene Regel, schon deswegen, weil die Meinungen der Parteien immer ihren Grund haben, und nach der Eingeschränktheit unserer Kenntnisse jeder Respekt verdient, und auch recht haben kann' (ibid., 467).
33. Eberhard, *Versuch*, does not mention the medium tradition.
34. Westfeld, *Erzeugung*, 12, criticises Euler's explanation of the colours of objects.
35. Lichtenberg, 'Schreiben'; Arbuthnot, 'Abhandlung'; Heinrich, 'Preisfrage'. Euler, *Briefe*, vol. 1, 201–46.
36. Ludwig, *Dissertatio*, 3–7; Werner, *Entwurf*, 80–1. Winkler, *Anfangsgründe* (1754), 275–86, also to a large extent bases his ideas on Euler's first chapter in the 'Nova theoria'. In Table 3 only Winkler, *Institutiones* (1738), is included.
37. Lambert, 'Instrumens'; see Section 3a, below. This article of Lambert's was published in 1770, ten years after the *Photometria*, and is therefore not included in Table 3.
38. Maler, *Physik*, 218–26; Suckow, *Entwurf*, 237; Eberhard, *Gründe* (1787), 444–6; Voigt, *Versuch*, 391–2, 396.
39. Euler to Lambert, 25 April 1961; Bopp, ed., 'Briefwechsel', 22.
40. Krafft, *Praelectiones*, vol. 3 (1754), 66–71 (survey), 71–7 (discussion of Euler's argumentation).
41. Ibid., 71–2, quotation 72: 'analogiae nimis sunt metaphysicae'.

42. Ibid., 73.
43. Ibid., 74.
44. Ibid., 66–7, 74–5.
45. Ibid., 63, 75.
46. Ibid., 76.
47. Ibid., 76–7. The article by Euler indicated here is his 'Explicatio'; see page 77.
48. An exception is the reference in Scheibel's bibliography, *Einleitung*, vol. 9 (1775), 375.
49. Mangold, *Systema*, 70–1.
50. Ibid., 71–2, quotation 72: 'Plurimi Recentiores'.
51. For example, ibid., 'Ad lectorem', [3]; 74, 88, 90.
52. Ibid., 80.
53. Schaff, *Geschichte*, 4–7, 154–9, 168, 170–1.
54. Mangold, *Systema*, 72–6. The last argument was, for example, also advanced by Wolff (see page 123).
55. Ibid., 80–1.
56. Ibid., 88–90.
57. Ibid., 91.
58. Ibid., 93–6.
59. Ibid., 98–101, especially 98: 'confusio'. In Euler, 'Pensées', Section 10, the unobstructed crossing of rays is actually discussed. Euler does not use the superposition principle; he prefers the argument that the distance between the pulses in the light rays is so great that the pulses do not obstruct each other. (N.B. Euler posits that the lowest frequency in the light is ten thousand vibrations a second.)
60. For the following, see pages 29–31, 54–6 (Newton, Huygens), 62–3 and 82–3 (Johann II Bernoulli), 133 (Euler), and (Mangold).
61. For the terms *negative analogy* and *positive analogy*, see Hesse, *Models*, 8.
62. For the adherents of an emission theory see pages 126–7 and the notes accompanying them. One example of someone who follows in Euler's footsteps is Ludwig, *Dissertatio*, 5–6.
63. Gabriel Cramer, in a letter to Euler in 1746, advances the idea that the difference in elasticity between the air and the light ether could explain the difference in propagation: In the extremely elastic ether, propagation in a straight line would be 'infinitely' easier than in the lateral directions. He says, a propos of this idea: 'Mr. Huygens semble avoir eu cette idée: mais je ne suis pas content de la manière dont il la develope' (Cramer to Euler, 30 August 1746, copy Euler archives, Basel, p. 22). Cramer also rejects Johann II Bernoulli's theory.
64. Segner, *Einleitung* (1754), 433–6, quotation 435.
65. Lambert, 'Instrumens', 87–90.
66. Ibid., 91 (emphasis in the original).
67. Ibid., 91–2, quotation 92.
68. Béguelin, 'Recherches' (1774).
69. Ibid., 160–1, quotation 160.
70. In the first half of the twentieth century, the first experiments with anechoic rooms were carried out; Trendelenburg, *Einführung*, 402–5.

71. Here Béguelin gives Newton's conception from the *Opticks* of 1704 (and earlier). The same conception was also customary in the textbooks of the emission tradition. As we have seen, Newton toned down his thesis, in the *Optice* of 1706, that sound waves penetrate *everywhere* into the shadow (see page 56).
72. Béguelin, 'Recherches' (1774), 161.
73. See pages 147 and 149. For Béguelin's philosophical position, see Hakfoort, "Béguelin", 299–300, and the literature given there.
74. Priestley, *History*, vol. 1, 357 (emphasis in the original).
75. Priestley, ed. Klügel, *Geschichte*, vol. 2, 264 (Klügel's addition).
76. Ibid., 261–2, quotation 262.
77. Ibid., 259, 262.
78. Ibid., 264, 553.
79. Gehler, *Wörterbuch*, vol. 2 (1789), 901–2, approvingly quotes Klügel's experiment but does not regard it as decisive: Both optical traditions have 'grosse Schwierigkeiten'.
80. Compare de Lang, "Originator", 21–7, who claims that a solution may also be reached by extending Huygens' approach, using *non*periodic waves and introducing the concept of destructive interference.
81. Priestley, *History*, vol. 2, 520–40.
82. Béguelin, 'Recherches' (1774), 156. Béguelin is obviously using Euler's conception of the propagation and refraction of light pulses. According to Euler, light pulses move in a strictly rectilinear fashion, and they only change direction with a change of medium. According to Huygens' pulse theory, however, secondary spherical pulses enter the glass cube. With an angle of incidence of 90°, these pulses do not concur to form a pulse front. Both medium theories thus predict that there will be no refraction where $i = 90°$.
83. Ibid., 156–7.
84. Ibid., 157.
85. Ibid., 156.
86. Béguelin's information on the experiment is not as detailed as could be desired. He devotes no consideration to the partial reflection at the upper side of the cube and elsewhere. Neither does he say in which point of the upper side the ray is refracted. Finally, he writes nothing about the decreasing intensity of the refracted ray where there is an increasing angle of incidence.
87. Ibid., 157. Béguelin also observed that refraction had already disappeared where the angle of incidence was slightly smaller than 90°. He considered this result to be incidental and of no importance for his conclusion that refraction is not caused by an attractive force (ibid., 157–8).
88. Ibid., 158.
89. Ibid., 158–60, quotations 159, 159–60.
90. Anon., Review of Béguelin.
91. Ibid., 21. The remark about Béguelin's use of the principle of sufficient reason is an allusion to Béguelin, 'Essais', and id., 'Application'.
92. Priestley, ed. Klügel, *Geschichte*, vol. 2, 264, 567.
93. Heinrich, 'Preisfrage', 259; Gehler, *Wörterbuch*, vol. 2 (1789), 902; Erxleben, ed. Lichtenberg, *Anfangsgründe* (1794), 278 (comment by Lichtenberg).
94. Béguelin, 'Recherches' (1779).

95. Senebier, 'Réponse', 202, 210–11. Marivetz, 'Réponse'; id., 'Lettre'.
96. Béguelin's article is not cited in Young, *Course*, or in the accompanying impressive bibliography; in Fresnel, *Oeuvres*, there is also no reference to the article. Biot, *Traité*, vol. 3, 256, 276.
97. Harvey, *History*, 306–7. Article 'Phosphorus' in Gehler, *Wörterbuch*, vol. 3 (1798), 475–85.
98. Harvey, *History*, 329–32; Wilson, *Series*.
99. Harvey, *History*, 314, 325, 333; Beccaria, 'Letter'.
100. Harvey, *History*, 361–2.
101. Euler, 'Nova theoria', 44 (see pages 112–13); id., *Lettres*, vol. 1 (1768), *Opera* edition vol. III.11, 63.
102. Priestley, *History*, vol. 1, 365 (emphasis added).
103. Wilson, *Series*, 22–4, 37, 76–7, quotations 78, 86 (emphasis in the original).
104. Perhaps partly misled by Priestley's placing of the emission tradition and the sponge hypothesis in opposition to the medium tradition, Cantor thinks (*Optics*, 60), incorrectly, that Wilson regarded the phenomena of phosphorescence as a decisive argument against the emission tradition and in favour of the medium tradition.
105. This result is not found in the first edition of Wilson's *Series*, but it is in the second edition of 1776; Wilson, 'Traduction', 93. I have been unable to consult the second edition. The result is also mentioned in Wilson's letter to the Berlin Academy; see the next note.
106. Formey, 'Sur la lumiere', 39. Secretary Formey reports in the History section of the Berlin memoirs on the discussion concerning Wilson's experiments. He does so by quoting literally from Wilson's letter to the Berlin Academy (36–38) and from the reports by Lambert (38), Marggraf (39), and Béguelin (39–44).
107. Béguelin's report is stated to have been included in its entirety (ibid., 39), which justifies the assumption that only part of Marggraf's and Lambert's reports were quoted.
108. Ibid., 39–40.
109. Ibid., 40–1, quotation 40.
110. Ibid., 41–2. See Béguelin, 'Remarques', especially 330.
111. Formey, 'Sur la lumiere', 43 (emphasis in the original).
112. Ibid., 44.
113. Ibid., 38 (emphasis in the original). This is all that is cited in Formey's reporting of Lambert. The sequence of the reports is clear from Formey, 'Ouvrages', 50–1.
114. Krafft, 'Expérience', 78–9.
115. Euler, 'Réflexions', 353, 355.
116. Wilson himself ('Traduction', 92, 97) protests against this version of matters, and maintains that his results indeed confirm Newton's theory of light and colours.
117. Euler, 'Réflexions', 353.
118. Ibid., 354–5, quotation 355.
119. Wilson, 'Traduction', 93–5, points out that phosphorescent light could be seen actually *during* the illumination (and not only after the phosphorus had

been taken into a dark space, as Euler thought). He thinks that this conflicts with Euler's explanatory mechanism.

120. Deluc, *Idées*, vol. 1, 123 (emphasis in the original).
121. Euler, 'Réflexions', 351–2.
122. Deluc, *Idées*, vol. 1, 123–5, quotations 124 (emphasis in the original), 123.
123. Magellan, 'Expérience', 154–5. W. L. Krafft, 'Expérience', 79, posits – in response to Allamand's experiments – that the idea of incident and emitted light having the same colour is only valid for certain kinds of phosphori, for example, the Bologna stone.
124. Gehler, *Wörterbuch*, vol. 3 (1798), 479–80. This is the second unaltered edition of Gehler's third volume (the first edition of this volume appeared between 1789 and 1795; see Section 6, below).
125. Priestley, *History*, vol. 1, 378–81; Eder, *Geschichte*, 26–7.
126. Priestley, *History*, vol. 1, 378.
127. Eder, *Geschichte*, 26–41; Morton, *History*, 328–36.
128. Senebier, *Mémoires*, vol. 1, 45–70; vol. 3, 198–205. Morton, *History*, 334–6; Eder, *Geschichte*, 34–7.
129. Eder, *Geschichte*, 39–41.
130. Gehler, *Wörterbuch*, vol. 2 (1789), 902; Fischer, *Geschichte*, vol. 7, 5.
131. Eder, *Geschichte*, 26 (note 3), 27 (note 1), 28 (note 2), 34 (note 1), 39 (note 2), 40 (note 1). The book by Ingen-Housz was translated into German the year after its publication; van der Pas, 'Ingen-Housz', 15.
132. Stichweh, *Entstehung*, especially 102–4; Hufbauer, *Formation*, especially 226–7; Schmauderer, 'Chemiatriker'; Schimank, 'Der Chemiker'; Weyer, 'Entwicklung'. For a general discussion of chemistry, physics, and optics, see Shapiro, *Fits*, 216–25.
133. Kraus, *Forschung*, 236–43, quotation 236.
134. Arbuthnot, 'Abhandlung'; Kraus, *Forschung*, 239.
135. Another example from the same period is Lichtenberg, 'Schreiben'.
136. Heinrich, 'Preisfrage'. For Heinrich, see Hartmann, 'Physiker'.
137. His most important work in this area is Heinrich, *Phosphorescenz*.
138. Heinrich, 'Preisfrage', 149–50.
139. Ibid., 153.
140. Ibid., 154.
141. Ibid., 167.
142. Ibid., 208. The original has 'meinem' instead of 'einem', but this is obviously a printer's error.
143. Ibid., 211–13, 217–18.
144. Ibid., 219–21, 225.
145. Ibid., 226, 247.
146. Ibid., 227–31.
147. Ibid., 231–5.
148. Ibid., 226–7. Heinrich nevertheless makes it clear that he prefers Lavoisier's theory; ibid., 226, 227, 238.
149. Ibid., 246–7, quotation 247.
150. Ibid., 248.
151. Ibid., 264–5.

152. Ibid., 294–9.
153. Ibid., 312–14, quotation 314.
154. Ibid., 320–2.
155. Heinrich's second chapter also contains supplementary arguments on the colours of opaque bodies, refraction (here Béguelin's experiment on grazing rays is also discussed; see page 150), and Euler's change of opinion on the question of whether red or violet has the greatest frequency (see page 105, above).
156. Kraus, *Forschung*, 240, appears to be issuing this reproach: '[D]ie Phänomene auf Grund von Vorstellungen Eulers zu erklären, prüfte er [Heinrich] nicht einmal'.
157. Voigt, *Versuch*, 383, only postulates that the sun's rays are active in photochemical phenomena 'sowohl durch ihre mechanischen Schütterungen als durch chemische Beymischung ihres Stoffs'.
158. Senebier, *Mémoires*, vol. 3, 245–6: 'Toutes les expériences que j'ai rapportées semblent concourir, pour faire voir dans la lumière un corps semblable à ceux que nous connoissons; elle a, comme eux, les mêmes propriétés, & sur-tout des affinités qui lui sont propres . . . ; c'est avoir fait un pas dans cette science [the physical and chemical investigation of light], que d'avoir prouvé que la lumière n'est pas seulement l'oscillation d'un fluide éthéré; mais qu'elle est véritablement un composé de petits corps, qui se combinent certainement avec des corps plus grands'.
159. Gehler, *Wörterbuch*, vol. 2 (1789), 902–3, quotations 902, 903.
160. Karsten, ed. Gren, *Anfangsgründe*, 541–3, quotations 541, 542.
161. Gren, *Grundriss* (1788), 298–9, quotation 298.
162. Erxleben, *Anfangsgründe* (1777), 244–5.
163. Id., ed. Lichtenberg, *Anfangsgründe* (1787), 253–333, especially 259–60; compare ibid. (1791), 267. I have not seen the first edition that was prepared for the press by Lichtenberg in 1784 (i.e., the third edition in the total). However, it is very unlikely that Lichtenberg would have adopted an anti-Euler position in 1784, only to do an about-face in 1787.
164. Lichtenberg, 'Schreiben'; Werner, *Entwurf*.
165. Erxleben, ed. Lichtenberg, *Anfangsgründe* (1791), 267.
166. Ibid.
167. Id., ed. Lichtenberg, *Anfangsgründe* (1794), 275.
168. Hoskin, 'Mining'.
169. Euler, ed. Kries, *Briefe*, vol. 1 (1792), 201–46.
170. Ibid., 203–10, 211–26.
171. Ibid., 226–39.
172. Ibid., 240–5, quotations xxiv, 241–2.
173. Ibid., 245.
174. Ibid., 245–6.
175. Gren, *Grundriss* (1793), 378.
176. Gehler, *Wörterbuch*, vol. 5 (1795), 546–9, quotation 547.
177. See also Ebermaier, *Versuch*, 40–1; Vieth, *Anfangsgründe*, 292; Fischer, *Geschichte*, vol. 7, 2–5.

6. *Epilogue: Optics as a mirror of eighteenth-century science*

1. Heilbron, 'Experimental', 360–1, 361–3, 383–4.
2. Kuhn, 'Mathematical'. This paper is based on a lecture presented in 1972 and was first published in 1975, in French. In 1976 it was slightly adapted for English publication; a year later it appeared unaltered in *The essential tension*.
3. Ibid., 31–2.
4. Ibid., 32–3.
5. Ibid., 33–4, quotation 33.
6. Ibid., 34.
7. Ibid., 32 (note) and 35 (note).
8. Ibid., 35–6, 46.
9. For example, Hall, 'Scholar'; Home, 'Straitjacket'; Hooykaas, 'Traditie'.
10. Kuhn, 'Mathematical', 35, 52, 63. Kuhn does not use the term 'second scientific revolution' in 'Mathematical', but he does so, in the same context, in an earlier article, 'Function', 218.
11. Heilbron, *Electricity*, xi, xv.
12. Home, 'Straitjacket', 244–8; id., Introduction to Aepinus, ed. Home, *Essay*, 193–224. It seems to follow from Home's account that he arrived independently of Kuhn at similar ideas. See, for example, his passing reference to Kuhn's article on the two traditions and his assertion that he offers a new view of eighteenth-century physics; id., 'Straitjacket', 244, 248.
13. Hall, *Revolution in science*, 15–19, 148–49. Hall's agreement with Kuhn's thesis is all the more striking because he does not consider Kuhn's ideas on paradigms to be applicable to the Scientific Revolution; ibid., 28–9, 344. Schuster, 'Scientific Revolution', 226.
14. Cohen, *Scientific Revolution*.
15. Kuhn, 'Function', 213–20, especially 215–16; id., 'Mathematical', especially 45–6, 48; also 36, 38, 41, 42, 43, 49–50, 56, 61–2. On p. 62 there is a hint of doubt about his way of categorising optics.
16. Cantor, *Optics*, 2–3, quotation 2, 3. Cantor adds that Kuhn's thesis on the two groups of physical sciences results in important questions for further research in the historiography of optics.
17. The first exposition of the three-fold division was in a lecture in 1985; see Hakfoort, 'Torn'. Since then, the scheme has been used and developed in id., 'Fundamentele'; id., 'Newton's "*Opticks*"'; id., 'Newton's optics'. For a discussion, see Cohen, *Scientific Revolution*.
18. Compare Stichweh, *Entstehung*, 14–39.
19. Wolff, *Anfangs-Gründe*; id., *Würckungen der Natur*; id., *Allerhand*. A fourth work in this series was on teleology in Nature: id., *Absichten*.
20. Harnack, *Geschichte*, vol. 1, 300.
21. Hakfoort, 'Newton's "*Opticks*"', 108; 'Newton's optics', 97–8. The uniqueness of Huygens' research style in seventeenth-century physical optics has been further defined and documented in Dijksterhuis, *Verhandelingen*.

    In Kuhn's view, only three people were able to combine in a significant way the mathematical and experimental traditions: Newton, Huygens, and Mariotte

(Kuhn, 'Mathematical', 50). The lack of synthesis of the *three* traditions in Newton's optical research is discussed in the text. *In optics* Mariotte has not done significant mathematical work.

Recently Shapiro has suggested that Newton's research on periodic colours is comparable to Huygens' work on double refraction: '[Newton's] explanation of the colors of thick plates is an unrecognized landmark in seventeenth-century mathematical physics in its combination of physical reasoning, mathematical theory, and quantitative experiment. . . . In optics only Huygens, with his explanation of double refraction by means of Huygens's principle, had achieved a comparable interplay of physical model, mathematical theory, and experiment'; 'The theory of fits, like Huygens's wave theory, began to be seriously studied and garner supporters only at the end of the eighteenth century just as the mathematical and quantitative style of physics, whose seeds were planted by Newton and Huygens, began to flourish' (Shapiro, *Fits*, 171, 203). Shapiro and I agree that Huygens' method in his research on double refraction was unique, or nearly unique, in optics before the end of the eighteenth century. However, we differ on the comparability of Newton's work with that of Huygens. As follows from Shapiro's fascinating account, Newton's research, as it was presented, did not involve the natural philosophical 'speculation' that Huygens freely used, because Newton scrupulously avoided to use hypotheses on the nature of light in the presentation of his theory of periodic colors. Of course, in his head, his manuscripts, and his published queries Newton did use hypotheses, but in print he presented a mathematical theory derived from observations that was allegedly free of hypotheses. Therefore, his theory is, in my terms, an example of the integration of two traditions: the mathematical and the experimental.

22. Shapiro, *Fits*, 171.
23. Haüy, 'Mémoire'; Buchwald, 'Investigations', 335–42.
24. Malebranche to P. Berrand, 1707; Malebranche, *Oeuvres*, vol. 19, 771–2.
25. For example, Karsten, *Naturlehre* (1780), 452, on the 'Nova theoria': 'Wer aber Herrn Euler hier folgen will, muss in der höhern Mathematik geübt seyn'.
26. Darjes, *Erste Gründe*, 603, 606.
27. Heilbron, *Electricity*, 1; id., 'Experimental', 373–5.
28. Frankel, 'Biot', especially 42–7, 59–64; Cantor, *Optics*, 61, 191; see also Whewell, *History*, vol. 2, 319–20. Shapiro, *Fits*, part 2, especially 211–25, 242–90, quotation x.
29. See also Hakfoort, 'Commentary', 26–8; compare Buchwald, *Rise*.
30. Kuhn, 'Mathematical', 64.
31. Ibid., 39, 43–4.
32. As before with Kuhn, in this case, too, there seems to be room for some discussion on his actual point of view. However, it is more important to look at the way in which Kuhn's thesis on the two scientific traditions are in fact used by him and others. There, the undervaluation of the natural philosophical element in early modern science is certainly present.

BIBLIOGRAPHY

Adikes, E., *Kant als Naturforscher* (2 vols.; Berlin, 1924–5).

Aepinus, F. U. T., ed. R. W. Home, *Essay on the theory of electricity and magnetism* (translation of *Tentamen theoriae electricitatis et magnetismi*, 1759; Princeton, 1979).

Albury, W. R., 'Halley and the *Traité de la lumière* of Huygens: New light on Halley's relationship with Newton', *Isis* 62(1971) 445–68.

Anon., Review of Béguelin, 'Recherches' (1774), *Allgemeine deutsche Bibliothek* 26(1775) 18–22.

Arbuthnot, B., 'Abhandlung über die Preisfrage von dem Eulerischen und Newtonischen Systeme vom Lichte, welcher eine goldene Medaille zuerkannt worden ist', *Neue philosophische Abhandlungen der Baierischen Akademie der Wissenschaften* 5(1789) 329–98.

Aristotle, 'De sensu', in *On the soul; Parva naturalia; On breath* (trans. W. S. Hett; London, 1957) 205–83.

Meteorologica (trans. H. D. P. Lee; London, 1962).

Badcock, A. W., 'Physical optics at the Royal Society 1660–1800', *The British Journal for the History of Science* 1(1962–3) 99–116.

Beccaria, G., 'Letter from Mr. John Baptist Beccaria, of Turin, F.R.S. to Mr. John Canton, F.R.S. on his new phosphorus receiving several colours, and only emitting the same', *Philosophical Transactions* 61(1771) 212.

Bechler, Z., 'Newton's search for a mechanistic model of colour dispersion: A suggested interpretation', *Archive for History of Exact Sciences* 11(1973) 1–37.

'Newton's law of forces which are inversely as the mass: A suggested interpretation of his later efforts to normalise a mechanistic model of optical dispersion', *Centaurus* 18(1974) 184–222.

Béguelin, N., 'Remarques et observations sur les couleurs prismatiques et sur l'analyse physique, et métaphysique, de la sensation des couleurs en général', *Histoire de l'academie royale des sciences et belles-lettres* 1764 (publ. 1766), Mémoires, 297–363.

'Recherches sur les moyens de découvrir par des expériences comment se fait la propagation de la lumiere', *Nouveaux mémoires de l'académie royale des sciences et belles-lettres* 1772 (publ. 1774), Mémoires, 152–62. (Rpt.: *Observations sur la physique, sur l'histoire naturelle et sur les arts* 13(1779) 38–46.)

'Essais sur un algorithme déduit du principe de la raison suffisante', *Nouveaux mémoires de l'académie royale des sciences et belles-lettres* 1772 (publ. 1774), Mémoires, 296–352.

'Application du principe de la raison suffisante à la démonstration d'un théoreme de M. Fermat sur les nombres polygonaux, qui n'a point encore demontré', *Nouveaux mémoires de l'académie royale des sciences et belles-lettres* 1772 (publ. 1774), Mémoires, 387–413.

Bernoulli, Johann I, 'Disquisitio catoptrico-dioptrica exhibens reflexionis & refractionis naturam, nova & genuina ratione ex aequilibrii fundamento deductam & stabilitam', *Acta eruditorum* 1701, 19–26.

Bernoulli, Johann II, 'Recherches physiques et géométriques sur la question: Comment se fait la propagation de la lumiére, proposée par l'académie royale des sciences pour le sujet du prix de l'année 1736', second separately paginated treatise in *Recueil des pieces qui ont remporté les prix de l'académie royale des sciences, depuis leur fondation jusqu'à présent; Avec les pieces qui y ont concouru*, vol. 3 (Paris, 1752).

Biot, J. B., *Traité de physique expérimentale et mathématique* (4 vols.; Paris, 1816).

Birch, T., *The history of the Royal Society of London for improving of natural knowledge, from its first rise* (4 vols., 1756–7; rpt. Hildesheim, 1968).

Blay, M., 'Une clarification dans le domaine de l'optique physique: *Bigness* et promptitude', *Revue d'histoire des sciences* 33(1980) 215–24.

Boeckmann, J. L., *Naturlehre; Oder: die gänzlich umgearbeitete Malerische Physik* (Karlsruhe, 1775).

Bopp, K., ed., 'Leonhard Eulers und Johann Heinrich Lamberts Briefwechsel', *Abhandlungen der Preussischen Akademie der Wissenschaften. Physikalisch-mathematische Klasse* 1924, no. 2, 1–45.

Bos, H. J. M., et al., eds., *Studies on Christiaan Huygens. Invited papers from the symposium on the life and work of Christiaan Huygens, Amsterdam, 22–25 August 1979* (Lisse, 1980).

Bošković, R. J., *Theoria philosophiae naturalis redacta ad unicam legem virium in natura existentium* (Venice, 1763).

A theory of natural philosophy (trans. J. M. Child; Cambridge, Mass., 1966).

Boyer, C. B., 'Early estimates of the velocity of light', *Isis* 33(1941–2) 24–40.

Broeckhoven, R. van, 'De kleurentheorie van Christiaan Huygens', *Scientiarum historia* 12(1970) 143–58.

Brunet, P., *L'introduction des théories de Newton en France au XVIIIe siècle avant 1738* (Paris, 1931).

Buchwald, J. Z., 'Experimental investigations of double refraction from Huygens to Malus', *Archive for History of Exact Sciences* 21(1979–80) 311–73.

The rise of the wave theory of light: Optical theory and experiment in the early nineteenth century (Chicago, 1989).

Burckhardt, J. J., et al., eds., *Leonhard Euler 1707–1783: Beiträge zu Leben und Werk; Gedenkband des Kantons Basel-Stadt* (Basel, 1983).

Caneva, K. L., 'From galvanism to electrodynamics: The transformation of German physics and its social context', *Historical Studies in the Physical Sciences* 9(1978) 63–159.

Cannon, J. T., and S. Dostrovsky, *The evolution of dynamics: Vibration theory from 1687 to 1742* (New York, 1981).

Canton, J., 'An easy method of making a phosphorus, that will imbibe and emit light, like the Bolognian stone; with experiments and observations', *Philosophical Transactions* 58(1768) 337–44.

Cantor, G. N., 'The changing role of Young's ether', *The British Journal for the History of Science* 5(1970–1) 44–62.

'The historiography of "Georgian" optics', *History of Science* 16(1978) 1–21.

Optics after Newton: Theories of light in Britain and Ireland, 1704–1840 (Manchester, 1983).

Cantor, G. N., and M. J. S. Hodge, eds., *Conceptions of ether: Studies in the history of ether theories 1740–1900* (Cambridge, 1981).

Cawood, J. A., 'The scientific work of D. F. J. Arago (1786–1853)' (Ph.D. diss., Leeds University, 1974).

Cohen, H. F., *The Scientific Revolution: An historiographical inquiry* (Chicago, 1994).

Cohen, I. B., 'Versions of Isaac Newton's first published paper', *Archives internationales d'histoire des sciences* 11(1958) 357–75.

'Isaac Newton, Hans Sloane and the Académie royale des sciences', in *L'aventure de la science: Mélanges Alexandre Koyré*, vol. 1 (Paris, 1964), 61–116.

The Newtonian revolution; With illustrations of the transformation of scientific ideas (Cambridge, 1980).

Dalham, F., *Institutiones physicae in usum nobilissimorum suorum auditorum adornatae* (3 vols.; Vienna, 1753–5).

Darjes, J. G., *Erste Gründe der gesamten Mathematik; Darinnen die Haupt-Theile so wohl der theoretischen als auch praktischen Mathematik in ihrer natürlichen Verknüpfung auf Verlangen und zum Gebrauch seiner Zuhörer entworfen* (Jena, 2nd ed. 1757).

Deluc, J. A., *Idées sur la météorologie* (2 vols.; London, 1786–7).

Desaguliers, J. T., 'An account of some experiments of light and colours, formerly made by Sir Isaac Newton, and mention'd in his Opticks, lately repeated before the Royal Society', *Philosophical Transactions* 29(1714–16) 433–47. (The article was published in no. 348, April–June 1716.)

'An account of an optical experiment made before the Royal Society, on Thursday, Dec. 6th, and repeated on the 13th, 1722', *Philosophical Transactions* 32(1722–3; publ. 1724) 206–8.

'Optical experiments made in the beginning of August 1728, before the president and several members of the Royal Society, and other gentlemen of several nations, upon occasion of signior Rizzetti's Opticks, with an account of the said book', *Philosophical Transactions* 35(1727–8) 596–629.

Descartes, R., *La dioptrique* (1637; edition used: *Oeuvres*, vol. 6, 81–228).

Les meteores (1637; edition used: *Oeuvres*, vol. 6, 229–366).

Principia philosophiae (1644; edition used: *Oeuvres*, vol. 8.1).

Oeuvres (11 vols.; Paris, 2nd ed. 1973–8).

Dijksterhuis, E. J., *De mechanisering van het wereldbeeld* (Amsterdam, 1950).

Dijksterhuis, F. J., 'Verhandelingen over het licht: De lichttheorie van Christiaan Huygens (1629–1695) en het ontstaan van de moderne fysica' (unpublished M.A. thesis, University of Twente, 1992).

Duhem, P., 'L'optique de Malebranche', *Revue de métaphysique et de morale* 1916, 37–91.

Dundon, S. J. S., 'Philosophical resistance to Newtonianism on the Continent, 1700–1760' (Ph.D. diss., St. John's University, 1972).
Eberhard, J. P., *Versuch einer näheren Erklärung von der Natur der Farben zur Erläuterung der Farbentheorie des Newton* (Halle, 1749).
Erste Gründe der Naturlehre (Halle, 1753, 5th ed. 1787).
Samlung derer ausgemachten Wahrheiten in der Naturlehre (Halle, 1755).
Ebermaier, J. C., *Versuch einer Geschichte des Lichtes in Rücksicht seines Einflusses auf die gesammte Natur, und auf den menschlichen Körper, ausser dem Gesichte, besonders* (Osnabrück, 1799).
Eder, J. M., *Geschichte der Photochemie und Photographie vom Alterthume bis in die Gegenwart* (Halle, 1891).
Elkana, Y., 'Newtonianism in the eighteenth century', *British Journal for the Philosophy of Science* 22(1971) 297–306.
Eneström, G., 'Der Briefwechsel zwischen Leonhard Euler und Johann I Bernoulli; II: 1736–1738', *Bibliotheca mathematica* III.5(1904) 248–91.
Verzeichnis der Schriften Leonhard Eulers (Jahresbericht der Deutschen Mathematiker-Vereinigung, Ergänzungsbände 4, 1. Lieferung; Leipzig, 1910).
'Bericht an die Eulerkommission der Schweizerischen naturforschenden Gesellschaft über die Eulerschen Manuskripte der Petersburger Akademie', *Jahresbericht der Deutschen Mathematiker-Vereinigung* 22(1913), 2. Abteilung, 191–205.
Ersch, J. S., *Handbuch der deutschen Literatur seit der Mitte des achtzehnten Jahrhunderts bis auf die neueste Zeit, systematisch bearbeitet und mit den nöthigen Registern versehen*, vol. III.2 (Leipzig, 1828; rpt.: *Handbuch etc.*, vol. 6, Hildesheim, 1982).
Erxleben, J. C. P., *Anfangsgründe der Naturlehre* (Göttingen, 2nd ed. 1777). (From the 3rd edition in 1784 the work was annotated by Lichtenberg.)
ed. G. C. Lichtenberg, *Anfangsgründe der Naturlehre* (Göttingen, 4th ed. 1787, 5th ed. 1791, 6th ed. 1794).
Euler, L., *Dissertatio physica de sono* (publ. 1727), in *Opera*, vol. III.1, 181–96.
'Pensées sur la lumiere et les couleurs', manuscript (1744) in the Archives of the Academy of Sciences, Leningrad: F.136, op. 1, no. 243, 1–25 (photocopy Euler Archive, Basel).
Opuscula varii argumenti, vol. 1 (Berlin, 1746).
'Nova theoria lucis et colorum' (publ. 1746), in *Opera*, vol. III.5, 1–45.
'Mémoire sur l'effet de la propagation successive de la lumière dans l'apparition tant des planètes que des comètes' (publ. 1746), in *Opera*, vol. III.5, 81–112.
'De relaxatione motus planetarum', in *Opuscula*, vol. 1 (1746), 245–76.
'Recherches physiques sur la nature des moindres parties de la matière' (publ. 1746), in *Opera*, vol. III.1, 6–15.
'Catalogus librorum meorum', manuscript (in or shortly after 1747) in the Archives of the Academy of Sciences, Leningrad: F.136, op. 1, no. 134, 383–402 (photocopy Euler Archive, Basel).
'Explicatio phaenomenorum quae a motu lucis successivo oriuntur' (publ. 1750), in *Opera*, vol. III.5, 46–80.
'Coniectura physica de propagatione soni ac luminis' (publ. 1750), in *Opera*, vol. III.5, 113–29.

'Essai d'une explication physique des couleurs engendrées sur des surfaces extrêmement minces' (publ. 1754), in *Opera*, vol. III.5, 156–71.

'Recherches physiques sur la diverse réfrangibilité des rayons de lumière' (publ. 1756), in *Opera*, vol. III.5, 218–38.

Anleitung zur Naturlehre, worin die Gründe zu Erklärung aller in der Natur sich ereignenden Begebenheiten und Veränderungen festgesetzet werden, in *Opera*, vol. III.1, 16–178.

'De la propagation du son' (publ. 1766), in *Opera*, vol. III.1, 428–51.

'Disquisitio de vera lege refractionis radiorum diversicolorum' (publ. 1768), in *Opera*, vol. III.5, 280–99.

Lettres à une princesse d'Allemagne sur divers sujets de physique & de philosophie (3 vols.; Petersburg, 1768–72; rpt.: *Opera*, vols. III.11–III.12).

'Réflexions sur quelques nouvelles expériences optiques, communiquées à l'académie des sciences, par Mr. Wilson' (publ. 1778), in *Opera*, vol. III.5, 351–5.

Oeuvres complètes en français, vol. 2 (Brussels, 1839).

Opera omnia (Basel, 1911–).

ed. J. H. S. Formey, 'Sur la lumiere et les couleurs', *Histoire de l'academie royale des sciences et des belles lettres de Berlin* 1745 (publ. 1746), Histoire, 17–24.

ed. F. C. Kries, *Briefe über verschiedene Gegenstände aus der Naturlehre* (3 vols.; Leipzig, 1792–4).

Fellmann, E. A., 'Leonhard Eulers Stellung in der Geschichte der Optik', in Burckhardt et al., eds., *Euler*, 303–30 (also in Euler, *Opera*, vol. III.9, 295–328).

Fischer, H., *Johann Jakob Scheuchzer (2. August 1672–23. Juni 1733). Naturforscher und Arzt* (Zurich, 1973).

Fischer, J. C., *Geschichte der Physik seit der Wiederherstellung der Künste und Wissenschaften bis auf die neuesten Zeiten* (8 vols.; Göttingen, 1801–8).

Fontenelle, B. de, 'Diverses observations de physique generale. IV', *Histoire de l'academie royale des sciences* 1720 (publ. 1722), Histoire, 11–12.

Formey, J. H. S., 'Sur la lumiere et les couleurs', *Nouveaux mémoires de l'académie royale des sciences et belles-lettres* 1776 (publ. 1779), Histoire, 36–44.

'Ouvrages imprimés ou manuscrits, machines et inventions, presentés à l'académie pendant le cours de l'année 1776', *Nouveaux mémoires de l'académie royale des sciences et belles-lettres* 1776 (publ. 1779), Histoire, 50–5.

Fox, R., *The caloric theory of gases from Lavoisier to Regnault* (Oxford, 1971).

Frankel, E., 'J. B. Biot and the mathematization of experimental physics in Napoleonic France', *Historical Studies in the Physical Sciences* 8(1977) 33–72.

Fresnel, A. J., *Oeuvres complètes* (3 vols.; Paris, 1866–70, rpt. New York, 1965).

Gabler, M., *Naturlehre; Zum Gebrauche öffentlicher Erklärungen* (Munich, 1780).

Gehler, J. S. T., *Physikalisches Wörterbuch oder Versuch einer Erklärung der vornehmsten Begriffe und Kunstwörter der Naturlehre mit kurzen Nachrichten von der Geschichte der Erfindungen und Beschreibungen der Werkzeuge begeleitet in alphabetischer Ordnung* (6 vols.; Leipzig, 1787–96, unaltered 2nd ed. 1798–1801).

Gerhardt, C. I., ed., *Briefwechsel zwischen Leibniz und Christian Wolff aus den*

Handschriften der Koeniglichen Bibliothek zu Hannover herausgegeben (Halle, 1860; rpt. Hildesheim, 1963).

Gillispie, C. C., ed., *Dictionary of scientific biography* (16 vols.; New York, 1970–80).

Gilson, É., *Études sur le role de la pensée médiévale dans la formation du système cartésien* (Paris, 2nd ed. 1951).

Goethe, J. W. von, *Beiträge zur Optik* (2 vols., 1791–2; edition used: *Schriften*, vol. I.3, 6–53).

Die Schriften zur Naturwissenschaft: Vollständige mit Erläuterungen versehene Ausgabe herausgegeben im Auftrage der Deutschen Akademie der Naturforscher Leopoldina (Weimar, 1947–).

Gordon, A., ed. B. Grant, *Physicae experimentalis elementa in usus academicos conscripta*, vol. 2 (Erfurt, 1753).

Grant, B., *Praelectiones encyclopaedicae in physicam experimentalem, et historiam naturalem* (Erfurt, 1770).

's Gravesande, W. J., *Physices elementa mathematica, experimentis confirmata; Sive introductio ad philosophiam Newtonianam* (2 vols.; Leyden, 1720–1, 3rd ed. 1742).

Philosophiae Newtonianae institutiones, in usus academicos (Leyden, 1728).

Gren, F. A. C., *Grundriss der Naturlehre zum Gebrauch akademischer Vorlesungen entworfen* (Halle, 1788).

Grundriss der Naturlehre in seinem mathematischen und chemischen Theile neu bearbeitet (Halle, 1793).

Grosses vollständiges Universal Lexicon aller Wissenschaften und Künste (64 vols., Halle, 1732–50; rpt. Graz, 1961–4).

Guerlac, H., *Essays and papers in the history of modern science* (Baltimore, 1977).

'Some areas for further Newtonian studies', *History of Science* 17(1979) 75–101.

Newton on the Continent (Ithaca, 1981).

Hakfoort, C., 'Wetenschapshistorische etiketten en de gevaren van polaire analyses', *Tijdschrift voor de geschiedenis der geneeskunde, natuurwetenschappen, wiskunde en techniek* 5(1982) 1–5.

'Christian Wolff tussen cartesianen en newtonianen', *Tijdschrift voor de geschiedenis der geneeskunde, natuurwetenschappen, wiskunde en techniek* 5 (1982), 27–38.

'Nicolas Béguelin and his search for a crucial experiment on the nature of light (1772)', *Annals of Science* 39(1982) 297–310.

'Torn between three lovers: On the historiography of 18th-century optics', in *XVIIth International Congress of History of Science, University of California, Berkeley, 31 July–8 August 1985, Acts*, vol. 1: *Abstracts of papers presented in scientific sections* (Berkeley, 1985), section Pc.

Optica in de eeuw van Euler: Opvattingen over de natuur van het licht, 1700–1795 (Amsterdam, 1986).

'De fundamentele spanning in Newtons natuurwetenschap', *Wijsgerig perspectief op maatschappij en wetenschap* 29(1988–9), no. 1, 2–7.

'Newton's "*Opticks*" and the incomplete Revolution', in P. B. Scheurer and G. Debrock, eds., *Newton's scientific and philosophical legacy* (Dordrecht, 1988), 99–112.

'Newton's optics: The changing spectrum of science', in J. Fauvel et al., eds., *Let Newton be*! (Oxford, 1988), 80–99.

'De schaduwzijde van Isaac Newton', *NRC Handelsblad*, section Wetenschap en Onderwijs, 11 June 1987.

'Naschrift', *NRC Handelsblad*, section Wetenschap en Onderwijs, 30 July 1987.

'Waarom waarderen we Huygens? Over het verschil tussen natuurkunde en geschiedenis van de natuurkunde', *Nederlands tijdschrift voor natuurkunde* A54(1988), no. 3/4, 88–92.

'Commentary on: J. Z. Buchwald, "The invention of polarization"', in R. P. W. Visser et al., eds., *New trends in the history of science. Proceedings of a conference held at the University of Utrecht* (Amsterdam, 1989), 23–8.

Halbertsma, K. T. A., *A history of the theory of colour* (Amsterdam, 1949).

Hall, A. R., 'The scholar and the craftsman in the scientific revolution', in M. Clagett, ed., *Critical problems in the history of science* (Madison, 1962), 3–23.

'Études Newtoniennes: IV. Newton's first book (I)', *Archives internationales d'histoire des sciences* 13(1960) 39–54.

'Newton in France: A new view', *History of Science* 13(1975) 233–50.

'Summary of the symposium', in Bos et al., eds., *Studies*, 302–13.

Philosophers at war: The quarrel between Newton and Leibniz (Cambridge, 1980).

'Further Newton correspondence', *Notes and Records of the Royal Society of London* 37(1982–3) 7–34.

The revolution in science, 1500–1750 (London, 1983).

All was light: An introduction to Newton's Opticks (Oxford, 1993).

Hamberger, A. A., *Kurzer Entwurf einer Naturlehre, worinnen alles aus dem einzigen Begriffe, das Kraft nichts anders als Druck sey, erwiesen ist* (Jena, 1781).

Hamberger, G. E., *Elementa physices, methodo mathematica in usum auditorii conscripta* (Jena, 1727, 5th ed. 1761).

Hankins, T. L., 'The reception of Newton's second law of motion in the eighteenth century', *Archives internationales d'histoire des sciences* 20(1967) 43–65.

Science and the Enlightenment (Cambridge, 1985).

Harnack, A., *Geschichte der Königlich preussischen Akademie der Wissenschaften zu Berlin* (3 vols.; Berlin, 1900).

Hartmann, L., 'Der Physiker und Astronom P. Placidus Heinrich von St. Emmeram in Regensburg (1758–1825)', *Studien und Mitteilungen zur Geschichte des Benediktiner-Ordens und seiner Zweige* 47(1929) 157–82, 316–51.

Harvey, E. N., *A history of luminescence from the earliest times until 1900* (Philadelphia, 1957).

Haüy, R.-J., 'Mémoire sur la double réfraction du spath d'Islande', *Histoire de l'académie royale des sciences* 1788, Mémoires, 34–61.

Heilbron, J. L., *Electricity in the 17th and 18th centuries: A study of early modern physics* (Berkeley, 1979).

'Experimental natural philosophy', in Rousseau et al., eds., *Ferment*, 357–87.

Heinrich, P., 'Ueber die Preisfrage: "Kömmt das Newtonische, oder das Eulerische System vom Lichte mit den neuesten Versuchen und Erfahrungen der Physik mehr überein?" Eine mit dem Preise belohnte Abhandlung', *Neue philosophische Abhandlungen der Baierischen Akademie der Wissenschaften* 5(1789) 145–328.

Die Phosphorescenz der Körper oder die im Dunklen bemerkbaren Lichtphänomene der anorganischen Natur, durch eine Reihe eigener Beobachtungen und Versuche geprüft und bestimmt (5 vols.; Nuremberg, 1811–20).

Hesse, M. B., *Models and analogies in science* (Notre Dame, 1966).

Hobert, J. P., *Grundriss des mathematischen und chemisch-mineralogischen Theils der Naturlehre* (Berlin, 1789).

Hollmann, S. C., *Primae physicae experimentalis lineae in auditorum gratiam ductae* (Göttingen, 1742).

Home, R. W., '"Newtonianism" and the theory of the magnet', *History of Science* 15(1977) 252–266.

'Out of a Newtonian straitjacket: Alternative approaches to eighteenth-century physical science', in *Studies in the eighteenth century, IV: Papers presented at the fourth David Nichol Smith memorial seminar Canberra 1976* (Canberra, 1979), 235–49.

'Newton on electricity and the aether', in Z. Bechler, ed., *Contemporary Newtonian research* (Dordrecht, 1982), 191–213.

'Leonhard Euler's "anti-Newtonian" theory of light', *Annals of Science* 45(1988) 521–33.

Hooykaas, R., 'De baconiaanse traditie in de natuurwetenschap', *Algemeen Nederlands tijdschrift voor wijsbegeerte en psychologie* 53(1961) 181–201.

Hoppe, E., *Die Philosophie Leonhard Eulers; Eine systematische Darstellung seiner philosophischen Leistungen* (Gotha, 1904).

Geschichte der Optik (1926; rpt. Wiesbaden, 1967).

Hoskin, M. A., '"Mining all within". Clarke's notes to Rohault's *Traité de physique*', *The Thomist* 24(1961) 353–63.

Hube, J. M., 'Betrachtungen über die Eindrücke, welche durch die Sinnen verursacht werden', *Hamburgisches Magazin, oder gesammlete Schriften, aus der Naturforschung und den angenehmen Wissenschaften überhaupt* 25, 2. Stück (1761) 184–98; 25, 4. Stück (1761) 353–70.

Hufbauer, K., *The formation of the German chemical community (1720–1795)* (Berkeley, 1982).

Huygens, C., *Traité de la lumiere; Où sont expliquées les causes de ce qui luy arrive dans la reflexion, & dans la refraction; Et particulierement dans l'etrange refraction du cristal d'Islande* (Leyden, 1690; rpt. Brussels, 1967).

Discours de la cause de la pesanteur (Leyden, 1690; appendix to *Traité*, 125–80).

Oeuvres complètes (22 vols.; The Hague, 1888–1950).

ed. W. J. 's Gravesande et al., *Opera reliqua* (2 vols.; Amsterdam, 1728).

Juškevič, A. P., 'Euler, Leonhard', in Gillispie, ed., *Dictionary*, vol. 4, 467–84.

and E. Winter, eds., *Leonhard Euler und Christian Goldbach, Briefwechsel 1729–1764* (Abhandlungen der Deutschen Akademie der Wissenschaften zu Berlin; Klasse für Philosophie, Geschichte, Staats-, Rechts- und Wirtschaftswissenschaften, 1965, no. 1; Berlin, 1965).

Kant, I., *Meditationum quarundam de igne succincta delineatio* (1755, publ. 1839),

in *Gesammelte Schriften, herausgegeben von der Königlich preussischen Akademie der Wissenschaften* (23 vols.; Berlin, 1900–5), vol. 1, 369–84.
Karsten, W. J. G., *Anfangsgründe der Naturlehre* (Halle, 1780).
ed. F. A. C. Gren, *Anfangsgründe der Naturlehre* (Halle, 1790).
Kaschube, J. W., *Cursus mathematicus, oder deutlicher Begrief der mathematischen Wissenschaften, mit gehörigen Rissen zur Bequehmlichkeit der Lehrenden und Lernenden auch ferneren Fortgang dieses nützlichen Studii mitgetheilet* (Jena, 2nd ed. 1718).
Elementa physicae mechanico-perceptivae, una cum appendice de geniis in augmentum scientiarum et usum studiosae iuventutis concinnata (Jena, 1718).
Kästner, A. G., 'Auszug aus Herrn Eulers Neuer Theorie des Lichts und der Farben, welche in dessen 1746 herausgekommenen *Opusculis varii argumenti* die dritte Stelle einnimmt', *Hamburgisches Magazin, oder gesammlete Schriften, zum Unterricht und Vergnügen, aus der Naturforschung und der angenehmen Wissenschaften überhaupt* 6, 2. Stück (1750) 156–97 (attributed to Kästner on the basis of Euler, *Opera*, vol. IVa.1, 199, 471).
Anfangsgründe der angewandten Mathematik (Göttingen, 2nd ed. 1765, 4th ed. 1792).
Kleinbaum, A. R., 'Jean Jacques Dortous de Mairan (1678–1771): A study of an Enlightenment scientist' (Ph.D. diss., Columbia University, 1970).
Klemm, F., *Die Geschichte der Emissionstheorie des Lichts* (Weimar, 1932).
Knorr, M., *Dissertatio dioptrica de refractione luminis* (Wittenberg, 1693).
Kopelevic, J. K., et al., eds., *Rukopisnye materialy Leonarda Eylera v arkhive Akademii nauk SSSR; Manuscripta Euleriana archivi Academiae scientiarum URSS* (2 vols.; Moscow, 1962–5).
Koyré, A., 'Études Newtoniennes: II. Les queries de l'Optique', *Archives internationales d'histoire des sciences* 13(1960) 15–29.
Krafft, F., 'Der Weg von den Physiken zur Physik an den deutschen Universitäten', *Berichte zur Wissenschaftsgeschichte* 1(1978) 123–62.
Krafft, G. W., *Praelectiones academicae publicae, commoda auditoribus methodo conscriptae* (Tübingen, 1750–4).
Krafft, W. L., 'Expérience sur le phosphore sulphuréo-calcaire de Mr. Canton', *Acta academiae scientiarum imperialis Petropolitanae* 1777 (publ. 1778), 77–9.
Kraus, A., *Die naturwissenschaftliche Forschung an der Bayerischen Akademie der Wissenschaften im Zeitalter der Aufklärung* (Munich, 1978).
Krüger, J. G., *Naturlehre*, vol. 1 (Halle, 3rd ed. 1750).
Kuhn, T. S., 'The caloric theory of adiabatic compression', *Isis* 49(1958) 132–40.
'The function of measurement in modern physical science' (publ. 1961), in *The essential tension*, 178–224.
The structure of scientific revolutions (Chicago, 1962).
'Mathematical versus experimental traditions in the development of physical science', in *The essential tension*, 31–65.
The essential tension: Selected studies in scientific tradition and change (Chicago, 1977).
Lambert, J. H., *Photometria sive de mensura et gradibus luminis, colorum et umbrae* (Augsburg, 1760).

'Sur quelques instrumens acoustiques', *Histoire de l'académie royale des sciences et belles-lettres* 19(1763; publ. 1770), 87–124.

Lang, H. de, 'Huygens', *NRC Handelsblad*, section Wetenschap en Onderwijs, 30 July 1987.

'Huygens' lichtvoortplantingstheorie geen golftheorie?', *Nederlands tijdschrift voor natuurkunde* A54(1988) 47–9.

'Weerwoord', *Nederlands tijdschrift voor natuurkunde* A54(1988) 92–93.

'Christiaan Huygens, originator of wave optics', in H. Blok et al., eds., *Huygens' principle 1690–1990: Theory and applications, Proceedings of an international symposium, The Hague/Scheveningen, November 19–22, 1990* (Amsterdam, 1992).

Lasswitz, K., *Geschichte der Atomistik vom Mittelalter bis Newton* (Hamburg, 1890; rpt.: 2 vols., Hildesheim, 1963).

Latchford, K. A., 'Thomas Young and the evolution of the interference principle' (Ph.D. diss., University of London, 1975).

Leibniz, G. W., *Opera omnia*, vol. 3 (Geneva, 1768).

ed. E. Gerland, *Nachgelassene Schriften physikalischen, mechanischen und technischen Inhalts* (Abhandlungen zur Geschichte der mathematischen Wissenschaften mit Einschluss ihrer Anwendungen begründet von Moritz Cantor, vol. 2; Leipzig, 1906).

Sämtliche Schriften und Briefe, herausgegeben von der Akademie der Wissenschaften der DDR, vol. III.1 (Berlin, 1976).

Lichtenberg, G. C., 'Schreiben an Herrn Werner in Giessen, die Newtonische Theorie vom Licht betreffend' (written in 1788 or 1789), in *Vermischte Schriften*, vol. 9 (Göttingen, 1806; rpt. Bern, 1972) 361–432.

Schriften und Briefe (4 vols.; Munich, 1967–72).

Lohne, J. A., 'Experimentum crucis', *Notes and Records of the Royal Society of London* 23(1968) 169–99.

Ludovici, C. G., *Ausführlicher Entwurff einer vollständigen Historie der Wolffischen Philosophie, zum Gebrauch seiner Zuhörer herausgegeben* (2 vols.; Leipzig, 2nd ed. 1737).

Ludwig, C., *Dissertatio de aethere varie moto causa diversitatis luminum* (Leipzig, 1773).

Mac Lean, J., 'De kleurentheorie van de aristotelianen en de opvattingen van De la Chambre, Duhamel en Vossius in de periode 1640–1670', *Scientarum historia* 10(1968) 208–25.

Magellan, J.-H., 'Sur une expérience faite avec le phosphore', *Observations sur la physique, sur l'histoire naturelle et sur les arts* 9(1777) 153–5.

Mairan, J. J. de, *Dissertation sur la cause de la lumiere des phosphores et des noctiluques* (Bordeaux, 1717).

'Recherches physico-mathematiques sur la reflexion des corps', *Histoire de l'academie royale des sciences* 1722 (publ. 1724), Mémoires, 6–51.

'Suite des recherches physico-mathematiques sur la reflexion des corps', *Histoire de l'academie royale des sciences* 1723 (publ. 1725), Mémoires, 343–86.

'Discours sur la propagation du son dans les différents tons qui le modifient', *Histoire de l'academie royale des sciences* 1737 (publ. 1740), Mémoires, 1–58 bis.

'Troisiéme partie des recherches physico-mathematiques sur la réflexion des corps', *Histoire de l'academie royale des sciences* 1738 (publ. 1740), Mémoires, 1–65.

'Quatriéme partie des recherches physico-mathématiques sur la réflexion des corps', *Histoire de l'academie royale des sciences* 1740 (publ. 1742), Mémoires, 1–58.

Malebranche, N., *Oeuvres* (22 vols.; Paris, 1958–84).

Maler, J. F., *Physik oder Naturlehre zum Gebrauch hoher und niederer Schulen* (Karlsruhe, 1767).

Mandelbaum, N. B., 'Jean-Paul Marat: The rebel as savant (1743–1788); A case study in careers and ideas at the end of the Enlightenment' (Ph.D. diss., Columbia University, 1977).

Mangold, J., *Systema luminis et colorum novam de refractione theoriam complectens cum praevia dissertatione de sono* (Ingolstadt, 1753).

Mariotte, E., *Oeuvres* (2 vols.; Leyden, 1717).

Marivetz, É. C. de, 'Réponse à M. Sennebier', *Observations sur la physique, sur l'histoire naturelle et sur les arts* 13, supplement (1778), 281–97.

'Lettre à M. Sennebier', *Observations sur la physique, sur l'histoire naturelle et sur les arts* 15(1780) 76–81.

Mautner, F. H., *Lichtenberg: Geschichte seines Geistes* (Berlin, 1968).

McGuire, J. E., and M. Tamny, *Certain philosophical questions: Newton's Trinity notebook* (Cambridge, 1983).

Melvill, T., 'Observations on light and colours', *Essays and Observations, Physical and Literary. Read before a Society in Edinburgh, and Published by them* 2(1756) 12–90.

Morton, A. G., *History of botanical science: An account of the development of botany from ancient times to the present day* (London, 1981).

Mouy, P., *Le développement de la physique cartésienne, 1646–1712* (Paris, 1934).

Murhard, F. W. A., *Litteratur der mathematischen Wissenschaften* (5 vols.; Leipzig, 1797–1805).

Musschenbroek, P. van, *Epitome elementorum physico-mathematicorum* (Leyden, 1726).

'Introductio ad cohaerentiam corporum firmorum', in *Physicae experimentalis et geometricae de magnete, tuborum capillarium vitreorumque speculorum attractione, magnitudine terrae, cohaerentia corporum firmorum dissertationes: ut et ephemerides meteorologicae Ultrajectinae* (Leyden, 1729), 421–672.

Elementa physicae conscripta in usus academicos (Leyden, 1734, 2nd ed. 1741).

Beginselen der natuurkunde, beschreven ten dienste der landgenooten (Leyden, 1736).

Beginsels der natuurkunde, beschreeven ten dienste der landgenooten (Leyden, 1739).

Grundlehren der Naturwissenschaft (trans. J. C. Gottsched; Leipzig, 1747).

Institutiones physicae conscriptae in usus academicos (2 vols.; Leyden, 1748).

Introductio ad philosophiam naturalem (2 vols.; Leyden, 1762).

Nagel, E., *The structure of science: Problems in the logic of scientific explanation* (London, 1974).

Newton, I., *Philosophiae naturalis principa mathematica* (London, 1687, Cambridge, 1713, London, 1726; edition used: variorum edition, ed. A. Koyré et al., 2 vols., Cambridge, 1972).

Opticks: Or, a treatise of the reflexions, refractions, inflexions and colours of light (London, 1704, rpt. Brussels, 1966; 2nd ed. 1717; 3rd ed. 1721; 4th ed. 1730, rpt. New York, 1952).

Optice: Sive de reflexionibus, refractionibus, inflexionibus & coloribus lucis (trans. S. Clarke; London, 1706).

'Remarks upon the observations made upon a chronological index of Sir Isaac Newton, translated into French by the observator and publish'd at Paris', *Philosophical Transactions* 33(1724–5) 315–21.

ed. I. B. Cohen, *Papers and letters on natural philosophy and related documents* (Cambridge, 1958).

ed. H. W. Turnbull et al., *The correspondence of Isaac Newton* (7 vols.; Cambridge, 1959–77).

ed. F. Cajori, *Mathematical principles of natural philosophy* (2 vols.; Berkeley, 1966).

ed. A. E. Shapiro, *The optical papers of Isaac Newton*, vol. 1, *The optical lectures 1670–1672* (Cambridge, 1984).

Pas, P. W. van der, 'Ingen-Housz, Jan', in Gillispie ed., *Dictionary*, vol. 7, 11–16.

Pater, C. de, *Petrus van Musschenbroek (1692–1761), een newtoniaans natuuronderzoeker* (Utrecht, 1979).

Pav, P. A., 'Eighteenth-century optics: The age of unenlightenment' (Ph.D. diss., Indiana University, 1964).

Planck, M., "Das Wesen des Lichts", in *Vorträge und Erinnerungen* (Stuttgart, 1749), 112–24.

Priestley, J., *The history and present state of discoveries relating to vision, light and colours* (2 vols.; London, 1772).

ed. G. S. Klügel, *Geschichte und gegenwärtiger Zustand der Optik, vorzüglich in Absicht auf den physikalischen Theil dieser Wissenschaft* (2 vols.; Leipzig, 1776).

Richter, G. F., 'De iis quae opticae Newtonianae in epistola ad Christinum Martinellum, non ita pridem opposuit Jo. Rizzettus brevis disquisitio', *Acta eruditorum; Supplementa* 8(1724) 226–34.

'Defensio disquisitionis suae contra Jo. Rizzettum', *Acta eruditorum* 44(1724) 27–39.

'De praecedente schediasmate Rizzettiano monitum', *Acta eruditorum; Supplementa* 8(1724) 488–93.

Rizzetti, G., 'De systemate opticae Newtonianae & de aberratione radiorum in humore crystallino refractorum; Excerpta e literis Joannis Rizzeti ad Christinum Martinellum, patricium Venetum', *Acta eruditorum; Supplementa* 8(1724) 127–42.

'Excerpta e novo exemplari epistolae seu dissertationis anti-Newtonianae', *Acta eruditorum; Supplementa* 8(1724) 234–6.

'Excerpta ex epistola Jo. Rizetti ad socios Societatis regiae Londinensis', *Acta eruditorum; Supplementa* 8(1724) 236–40.

'Super disquisitionem G. Frid. Richteri, de iis quae opticae Newtonianae Joh.

Rizzettus opposuit, responsio', *Acta eruditorum; Supplementa* 8(1724) 303–19.

'Ad responsionem, quam Jo. Rizzettus contra opticam Newtonianam dedit G. Frid. Richtero appendix', *Acta eruditorum; Supplementa* 8(1724) 394–8.

'Responsio ad G.Fr. Richterum, de optica Newtoniana', *Acta eruditorum; Supplementa* 8(1724) 484–8.

Saggio dell'antinewtonianismo sopra del moto e dei colori (Venice, 1741).

Robinet, A., *Malebranche de l'académie des sciences: L'oeuvre scientifique, 1674–1715* (Paris, 1970).

Ronchi, V., *Histoire de la lumière* (Paris, 1956).

Rosenberger, F., *Die Geschichte der Physik in Grundzügen mit synchronistischen Tabellen der Mathematik, der Chemie und beschreibenden Naturwissenschaften sowie der allgemeinen Geschichte* (3 vols., 1882–90; rpt.: 2 vols., Hildesheim, 1965).

Rosenfeld, L., 'Le premier conflit entre la théorie ondulatoire et la théorie corpusculaire de la lumière', *Isis* 11(1928) 111–22.

Rousseau, G. S., et al., eds., *The ferment of knowledge: Studies in the historiography of eighteenth-century science* (Cambridge, 1980).

Sabra, A. I., 'Newton and the "bigness" of vibrations', *Isis* 54(1963) 267–8.

Theories of light from Descartes to Newton (Cambridge, 1967, 2nd ed. 1981).

Schaff, J., *Geschichte der Physik an der Universität Ingolstadt* (Erlangen, 1912).

Schagrin, M. L., 'Experiments on the pressure of light in the 18th century', *Studia Leibnitiana supplementa* 13(1974) 217–39.

Scheibel, J. E., *Einleitung zur mathematischen Büchererkentnis* (16 vols.; Breslau 1769–86).

Scheuchzer, J. J., *Physica, oder Natur-Wissenschaft* (2 vols.; 1701, 2nd ed. 1711, 3rd ed. 1729).

Schimank, H., 'Zur Geschichte der Physik an der Universität Göttingen vor Wilhelm Weber (1734–1830)', *Rete* 2(1973–5) 207–52.

'Der Chemiker im Zeitalter der Aufklärung und des Empire (1720–1820)', in Schmauderer, ed., *Der Chemiker*, 207–58.

Schlosser, E. O., *Die Rezensionstätigkeit von Leibniz auf mathematischem und physikalischem Gebiet* (Bottrop, 1934).

Schmaling, L. C., *Naturlehre für Schulen* (Göttingen, 1774).

Schmauderer, E., 'Chemiatriker, Scheidekünstler und Chemisten des Barock und der frühen Aufklärungszeit', in Schmauderer, ed., *Der Chemiker*, 101–205.

ed., *Der Chemiker im Wandel der Zeiten: Skizzen zur geschichtlichen Entwicklung des Berufsbildes* (Weinheim, 1973).

Schmid, N., *Von den Weltkörpern; Zur gemeinnützigen Kenntnis der grossen Werke Gottes* (Leipzig, 2nd ed. 1772).

Schofield, R. E., *Mechanism and materialism: British natural philosophy in an Age of Reason* (Princeton, 1970).

Schuster, J. A., 'The Scientific Revolution', in R. C. Olby et al., eds., *Companion to the history of modern science* (London, 1990), 217–42.

Segner, J. A., *De raritate luminis quibusdam praemissis* (Göttingen, 1740).

Einleitung in die Natur-Lehre (Göttingen, 1746, 2nd ed. 1754).

Senebier, J., 'Réponse à la lettre de Madame de V*** [= Marivetz]', *Observations sur la physique, sur l'histoire naturelle et sur les arts* 14(1779) 200–15.

Mémoires physico-chymiques, sur l'influence de la lumière solaire pour modifier les êtres des trois règnes de la nature, & sur-tout ceux du regne végétal (3 vols.; Geneva, 1782).

Shapiro, A. E., 'Rays and waves: A study in seventeenth century optics' (Ph.D. diss., Yale University, 1970).

'Kinematic optics: A study of the wave theory of light in the seventeenth century', *Archive for history of exact sciences* 11(1973) 134–266.

'Light, pressure and rectilinear propagation: Descartes' celestial optics and Newton's hydrostatics', *Studies in History and Philosophy of Science* 5(1974) 239–96.

'Huygens' kinematic theory of light', in Bos et al., eds., *Studies*, 200–20.

Fits, passions, and paroxyms: Physics, method, and chemistry and Newton's theories of colored bodies and fits of easy reflection (Cambridge, 1993).

Silliman, R. H., 'Augustin Fresnel (1788–1827) and the establishment of the wave theory of light' (Ph.D. diss., Princeton University, 1968).

'Fresnel and the emergence of physics as a discipline', *Historical Studies in the Physical Sciences* 4(1974) 137–62.

Speiser, D., 'Einleitung', in Euler, *Opera*, vol. III.5, vii–lviii.

'L'oeuvre de L. Euler en optique physique', in *Roemer et la vitesse de la lumière: Table ronde du Centre national de la recherche scientifique; Paris 16 et 17 juin 1976* (Paris, 1978), 207–24.

'Eulers Schriften zur Optik, zur Elektrizität und zum Magnetismus', in Burckhardt et al., eds., *Euler*, 215–28.

Steffens, H. J., *The development of Newtonian optics in England* (New York, 1977).

Steiger, R., 'Verzeichnis des wissenschaftlichen Nachlasses von Johann Jakob Scheuchzer (1672–1733)', *Beiblatt zur Vierteljahrsschrift der Naturforschenden Gesellschaft in Zürich* 21(1978) 1–75.

Stichweh, R., *Zur Entstehung des modernen Systems wissenschaftlicher Disziplinen: Physik in Deutschland 1740–1890* (Frankfurt, 1984).

Suckow, L. J. D., *Entwurf einer Naturlehre* (Jena, 2nd ed. 1782).

Thackray, A., '"Matter in a nut-shell": Newton's *Opticks* and eighteenth-century chemistry', *Ambix* 15(1968) 29–53.

Atoms and powers: An essay in Newtonian matter-theory and the development of chemistry (Cambridge, Mass., 1970).

Thiele, R., *Leonhard Euler* (Leipzig, 1982).

Trendelenburg, F., *Einführung in die Akustik* (Berlin, 1961).

Truesdell, C. A., 'Rational fluid mechanics, 1687–1765', in Euler, *Opera*, vol. II.12, vii–cxxv.

'Editor's introduction', in Euler, *Opera*, vol. II.13, vii–cxviii.

The rational mechanics of flexible or elastic bodies 1638–1788; Introduction to Leonhardi Euleri opera omnia vol. X et XI seriei secundae [= Euler, *Opera*, vol. II.11.2].

Essays in the history of mechanics (Berlin, 1968).

Vieth, G. U. A., *Anfangsgründe der Naturlehre für Bürgerschulen* (Leipzig, 1801).
Voigt, J. H., *Versuch einer neuen Theorie des Feuers, der Verbrennung, der künstlichen Luftarten, des Athmens, der Gährung, der Electricität, der Meteoren, des Lichts und des Magnetismus; Aus Analogien hergeleitet und durch Versuche bestätiget* (Jena, 1793).
Weidler, J. F., *Experimentorum Newtonianorum de coloribus explicationum novam veteri hypothesi accommodatam dissertatione optica* (Wittenberg, 1720).
Institutiones matheseos selectis observationibus illustratae in usum praelectionum academicarum (Wittenberg, 3rd ed. 1736).
Weinmann, K. F., *Die Natur des Lichts: Einbeziehung eines physikgeschichtlichen Themas in den Physikunterricht* (Darmstadt, 1980).
Werner, G. F., *Entwurf einer neuen Theorie der anziehenden Kräfte, des Aethers, der Wärme und des Lichts* (Frankfurt, 1788).
'Versuch einer Theorie des Ethers', *Magazin für das Neueste aus der Physik und Naturgeschichte* 5(1788) part 2, 95–108.
Westfall, R. S., 'The development of Newton's theory of color', *Isis* 53(1962) 339–58.
'Newton and his critics on the nature of colors', *Archives internationales d'histoire des sciences* 15(1962) 47–58.
Never at rest: A biography of Isaac Newton (Cambridge, 1980).
Westfeld, C. F. G., *Die Erzeugung der Farben: Eine Hypothese* (Göttingen, 1767).
Weyer, J., 'Die Entwicklung der Chemie zu einer Wissenschaft zwischen 1540 und 1740', *Berichte zur Wissenschaftsgeschichte* 1(1978) 113–21.
Whewell, W., *History of the inductive sciences* (3 vols., London, 1837, 3rd ed. 1857; rpt., *The historical and philosophical works*, vols. 2–4, London 1967).
Whittaker, E., *A history of the theories of aether and electricity* (2 vols.; New York, 1960).
Wiener, O., 'Der Wettstreit der Newtonschen und Huygensschen Gedanken in der Optik', *Berichte über die Verhandlungen der Sächsischen Akademie der Wissenschaften zu Leipzig, mathematisch-physische Klasse* 71(1919) 240–54.
Wilde, E., *Geschichte der Optik, vom Ursprunge dieser Wissenschaft bis auf die gegenwärtige Zeit* (2 vols. 1838–43; rpt. Wiesbaden 1968).
Wilson, B., *A series of experiments relating to phosphori and the prismatic colours they are found to exhibit in the dark* (London, 1775).
'Traduction d'une lettre de M. Wilson, adressée à M. Euler. Lue à la Societé Royale, le . . . de Juin 1779', *Observations sur la physique, sur l'histoire naturelle et sur les arts* 15(1780) 92–8.
Winkler, J. H., *Institutiones mathematico-physicae experimentis confirmatae* (Leipzig, 1738).
Anfangsgründe der Physic (Leipzig, 2nd ed. 1754).
Winter, E., ed., *Die Registres der Berliner Akademie der Wissenschaften 1746–1766: Dokumente für das Wirken Leonhard Eulers in Berlin* (Berlin, 1957).
Wolf, A., *A history of science, technology, and philosophy in the eighteenth century* (New York, 1939).
Wolff, C., Review of Newton, *Opticks* (1704), *Acta eruditorum* 1706, 59–64 (attributed to Wolff in Ludovici, *Entwurff*, vol. 2, 186).

Anfangs-Gründe aller mathematischen Wissenschaften (4 vols.; Halle, 1710).
Review of J. Rohault, ed. S. Clarke, *Physica* (London, 3rd ed. 1710), *Acta eruditorum* 1713, 444–8 (attributed to Wolff in Ludovici, *Entwurff*, vol. 2, 210).
Elementa matheseos universae (2 vols.; Halle, 1713–15).
'Notanda circa theoriam colorum Newtonianam', *Acta eruditorum* 1717, 232–4 (attributed to Wolff in Ludovici, *Entwurff*, vol. 2, 164).
Vernünfftige Gedancken von den Würckungen der Natur, den Liebhabern der Wahrheit mitgetheilet (Halle, 1723; rpt., *Gesammelte Werke*, vol. I.6, Hildesheim, 1981).
Vernünfftige Gedancken von den Absichten der natürlichen Dinge, den Liebhabern der Wahrheit mitgetheilet (2nd ed., Frankfurt, 1726; rpt., *Gesammelte Werke*, vol. I.7, Hildesheim, 1980).
Allerhand nützliche Versuche, dadurch zu genauer Erkäntnis der Natur und Kunst der Weg gebähnet wird, denen Liebhabern der Wahrheit mitgetheilet (3 vols.; Halle, 2nd ed. 1727–9; rpt., *Gesammelte Werke*, vols. I.20.1–I.20.3, Hildesheim, 1982).
Young, T., 'On the theory of light and colours', *Philosophical Transactions* 92(1802) 12–48.
'Experiments and calculations relative to physical optics', *Philosophical Transactions*, 94(1804) 1–16.
A course of lectures on natural philosophy and the mechanical arts (2 vols.; London, 1807).
Zeiher, J. E., *Abhandlung von denjenigen Glasarten, welche eine verschiedene Kraft die Farben zu zerstreuen besitzen* (Petersburg, 1763).
Zieglerin, J. C., *Grundriss einer natürlichen Historie und eigentlichen Naturlehre für das Frauenzimmer* (Halle, 1751).

INDEX